"Only a writer in love with his subject could have produced so charming a narrative about a bridge. There are many stories within the story of *The Bridge*. All are worth reading."
—*Houston Post*

"Talese has spun a fascinating, engrossing account of the construction of the Verrazano-Narrows Bridge. This is an absorbing drama; superbly written."
—*Times Union* (Jacksonville)

"No finer tribute in print will ever be found than this book."
—*Wilmington News Journal*

"Talese tells warm, funny and tragic stories of men, women, steel and concrete. This book is fine reading."
—*Denver Post*

"Fine writing and story-telling. . . . Superbly well does Talese tell his story, one that combines sadness, high humor, bawdiness, danger, death and poignancy in one fine package that readers will find hard to put down."
—*Arizona Republic*

"Talese is a shining example for all writers. He gets the drift of the story. . . . A complete, informative and fascinating account of the bridge."
—*Times* (Indianapolis)

THE BRIDGE

WALKER & COMPANY

NEW YORK

GAY TALESE

THE BRIDGE

To the ironworkers—
especially Gerard McKee and Danny Montour

First published in the United States of America in 2003 by Walker
Publishing Company, Inc.

Published simultaneously in Canada by Fitzhenry and Whiteside,
Markham, Ontario L3R 4T8

For information about permission to reproduce selections from this
book, write to Permissions, Walker & Company, 435 Hudson Street,
New York, New York 10014

Original captions for Lili Réthi's on-site illustrations by H. George
Decancq, Resident Engineer, Field Office, Ammann and Whitney

Library of Congress Cataloging-in-Publishing Data
available upon request
ISBN 0-8027-7644-2

Book design by Maura Fadden Rosenthal/*mspace*

Visit Walker & Company's Web site at www.walkerbooks.com

Printed in the United States of America
2   4   6   8   10   9   7   5   3   1

C O N T E N T S

P R E F A C E

A great bridge is a poetic construction of enduring beauty and utility, and in the early 1960s I would often don a hard hat and follow the workers across the catwalk of the Verrazano-Narrows Bridge and watch for hours as they crawled like spiders up and down the cable ropes and straddled beams while tightening bolts with their spud wrenches; and sometimes they would give a shove with their gloved hands against a stalled spinning wheel, or would bang a shoulder against a two thousand-pound piece of framework dangling from a crane—the framework representing one of millions of links in the rainbow-shaped roadway that would extend for two and a half miles horizontally across the Upper Bay of New York to connect the boroughs of Brooklyn and Staten Island, spreading discontent and trepidation among residents of both places.

I was less interested in the mundane matters of expanded land development and increased auto traffic and the potential of unwanted new neighbors than I was in observing the process by which artistry is achieved in the air through the fusion of drawing-board ingenuity and steely nerved bridge builders, both groups of men leaving a lasting impression on the skyline of New York and enhancing its spirit of grandiosity. The Verrazano-Narrows Bridge, the longest suspension span in the nation, is dominated by two towers each seventy stories high, and from these vantage points one can survey the panorama of the city—the Empire State Building, the Chrysler Building, the venerable Brooklyn Bridge, completed in 1883, the spires of Wall Street, and, until September 11, 2001, the 110-story twin towers of the World Trade Center.

When I first moved to New York, a half-century ago, I often asked myself: Whose fingerprints are on the bolts and beams of these soaring edifices in this overreaching city? Who are the high-wire walkers wearing boots and hard hats, earning their living by risking their lives in places where falls are often fatal and where the bridges and skyscrapers are looked upon as sepulchers by the families and coworkers of the deceased? We often know the names of the architects or chief engineers of renowned structures, but those men whose job it is to ascend to high and dangerous places—the kind of men who erected and connected the steel on the Empire State Building or spun the cables across the Brooklyn Bridge—are not identified by name in the books, archival materials, or other written accounts concerned with the construction of these landmarks.

I kept this in mind when I decided, in 1962, to write about the Verrazano-Narrows Bridge construction; it would include the names and biographical information about the workers, establishing their rightful place in the history of this grand undertaking. Now I am particularly gratified that nearly forty years after the hardcover publication of *The Bridge* in 1964, a paperback edition is being made available. Several of the individuals featured in the original work are dead, while other hard-hatted veterans of the Verrazano are not only alive but are still earning their living on high steel. I have included additional information about these men and other individuals who were involved with the story in an afterword following this edition's final chapter.

THE BRIDGE

THE BOOMERS

They drive into town in big cars, and live in furnished rooms, and drink whiskey with beer chasers, and chase women they will soon forget. They linger only a little while, only until they have built the bridge; then they are off again to another town, another bridge, linking everything but their lives.

They possess none of the foundation of their bridges. They are part circus, part gypsy—graceful in the air, restless on the ground; it is as if the wide-open road below lacks for them the clear direction of an eight-inch beam stretching across the sky six hundred feet above the sea.

When there are no bridges to be built, they will build skyscrapers, or highways, or power dams, or anything that promises a challenge—and overtime. They will go anywhere, will drive a thousand miles all day and night to be part of a new building boom.

They find boom towns irresistible. That is why they are called "the boomers."

In appearance, boomers usually are big men, or if not always big, always strong, and their skin is ruddy from all the sun and wind. Some who heat rivets have charred complexions; some who drive rivets are hard of hearing; some who catch rivets in small metal cones have blisters and body burns marking each miss; some who do welding see flashes at night while they sleep. Those who connect steel have deep scars along their shins from climbing columns. Many boomers have mangled hands and fingers sliced off by slipped steel. Most have taken falls and broken a limb or two. All have seen death.

They are cocky men, men of great pride, and at night they brag and build bridges in bars, and sometimes when they are turning to leave, the bartender will yell after them, "Hey, you guys, how's about clearing some steel out of here?"

Stray women are drawn to them, like them because they have money and no wives within miles—they liked them well enough to have floated a bordello boat beneath one bridge near St. Louis, and to have used upturned hardhats for flowerpots in the red-light district of Paducah.

On weekends some boomers drive hundreds of miles to visit their families, are tender and tolerant, and will deny to the heavens any suggestion that they raise hell on the job—except they'll admit it in whispers, half proud, half ashamed, fearful the wives will hear and then any semblance of marital stability will be shattered.

Like most men, the boomer wants it both ways. Occasionally his family will follow him, living in small hotels or trailer courts, but it is no life for a wife and child.

The boomer's child might live in forty states and attend a

dozen high schools before he graduates, if he graduates, and though the father swears he wants no boomer for a son, he usually gets one. He gets one, possibly, because he really wanted one, and maybe that is why boomers brag so much at home on weekends, creating a wondrous world with whiskey words, a world no son can resist because this world seems to have everything: adventure, big cars, big money—sometimes $350 or $450 a week—and gambling on rainy days when the bridge is slippery, and booming around the country with Indians who are sure-footed as spiders, with Newfoundlanders as shifty as the sea they come from, with roaming Rebel riveters escaping the poverty of their small Southern towns, all of them building something big and permanent, something that can be revisited years later and pointed to and said of: "See that bridge over there, son—well one day, when I was younger, I drove twelve hundred rivets into that goddamned thing."

They tell their sons the good parts, forgetting the bad, hardly ever describing how men sometimes freeze with fear on high steel and clutch to beams with closed eyes, or admitting that when they climb down they need three drinks to settle their nerves; no, they dwell on the glory, the overtime, not the weeks of unemployment; they recall how they helped build the Golden Gate and Empire State, and how their fathers before them worked on the Williamsburg Bridge in 1902, lifting steel beams with derricks pulled by horses.

They make their world sound as if it were an extension of the Wild West, which in a way it is, with boomers today still regarding themselves as pioneering men, the last of America's unhenpecked heroes, but there are probably only a thousand of them left who are footloose enough to go anywhere to build anything. And when they arrive at the newest boom town, they hold brief reunions in bars, and talk about old times, old faces: about Cicero Mike, who once drove a Capone whiskey truck during Prohibition and recently fell to his death off a bridge near Chicago; and Indian Al Deal, who kept three women happy out West and came to the bridge each morning in a fancy silk shirt; and about Riphorn Red, who used to paste twenty-dollar bills along the sides of his suitcase and who went berserk one night in a cemetery. And there was the Nutley Kid, who smoked long Italian cigars and chewed snuff and used toilet water and, at lunch, would drink milk and beer—without taking out the snuff. And there was Ice Water Charley, who on freezing wintry days up on the bridge would send apprentice boys all the way down to fetch hot water, but by the time they'd climbed back up, the water was cold, and he would spit it out, screaming angrily, *"Ice water, ice water!"* and send them all the way down for more. And there was that one-legged lecher, Whitey Howard, who, on a rail bridge one day, did not hear the train coming, and so he had to jump the tracks at the last second, holding on to the edge, during which time his wooden left leg fell off, and Whitey spent the rest of his life bragging about how he lost his left leg twice.

Sometimes they go on and on this way, drinking and reminiscing about the undramatic little things involving people known only to boomers, people seen only at a distance by the rest of the world, and then they'll start a card game, the first of hundreds to be played in this boom town while the bridge is being built—a bridge many boomers will never cross. For before the bridge is finished, maybe six months before it is opened to traffic, some boomers get itchy and want to move elsewhere. The challenge is dying. So is the overtime. And they begin to wonder: "Where next?" This is what they were asking one another in the early spring of 1957, but some boomers already had the answer: New York.

New York was planning a number of bridges. Several projects were scheduled upstate, and New York City alone, between 1958
and 1964, planned to spend nearly $600,000,000 for, among other
things, the double-decking of the George Washington Bridge, the
construction of the Throgs Neck Bridge across Long Island Sound—
and, finally, in what might be the most challenging task of a
boomer's lifetime, the construction of the world's largest suspension
span, the Verrazano-Narrows Bridge.

The Verrazano-Narrows, linking Brooklyn and Staten Island
(over the futile objections of thousands of citizens in both boroughs),
would possess a 4,260-foot center span that would surpass San
Francisco's Golden Gate by sixty feet, and would be 460 feet longer
than the Mackinac Bridge in upper Michigan, just below Canada.

It was the Mackinac Bridge, slicing down between Lake
Huron and Lake Michigan and connecting the cities of St. Ignace
and Mackinaw City, that had attracted the boomers between the
years 1954 and 1957. And though they would now abandon it for
New York, not being able to resist the big movement eastward, there
were a few boomers who actually were sorry to be leaving Michigan,
for in their history of hell-raising there never had been a more bom-
bastic little boom town than the once tranquil St. Ignace.

Before the boomers had infiltrated it, St. Ignace was a
rather sober city of about 2,500 residents, who went hunting in
winter, fishing in summer, ran small shops that catered to tourists,
helped run the ferryboats across five miles of water to Mackinaw
City, and gave the local police very little trouble. The land had been
inhabited first by peaceful Indians, then by French bushrangers,
then by missionaries and fur traders, and in 1954 it was still clean
and uncorrupt, still with one hotel, called the Nicolet—named after a white man, Jean Nicolet, who in 1634 is said to have paddled

in a canoe through the Straits of Mackinac and discovered Lake Michigan.

So it was the Nicolet Hotel, and principally its bar, that became the boomers' headquarters, and soon the place was a smoky scene of nightly parties and brawls, and there were girls down from Canada and up from Detroit, and there were crap games along the floor—and if St. Ignace had not been such a friendly city, all the boomers might have gone to jail and the bridge might never have been finished.

But the people of St. Ignace were pleased with the big new bridge going up. They could see how hard the men worked on it and they did not want to spoil their little fun at night. The merchants, of course, were favorably disposed because, suddenly, in this small Michigan town by the sea, the sidewalks were enhanced by six hundred or seven hundred men, each earning between $300 and $500 a week—and some spending it as fast as they were making it.

The local police did not want to seem inhospitable, either, and so they did not raid the poker or crap games. The only raid in memory was led by some Michigan state troopers; and when they broke in, they discovered gambling among the boomers another state trooper. The only person arrested was the boomer who had been winning the most. And since his earnings were confiscated, he was unable to pay the $100 fine and therefore had to go to jail. Later that night, however, he got a poker game going in his cell, won $100, and bought his way out of jail. He was on the bridge promptly for work the next morning.

It is perhaps a slight exaggeration to suggest that, excepting state troopers, everybody else in St. Ignace either fawned upon or quietly tolerated boomers. For there were some families who forbade their daughters to date boomers, with some success, and there were young local men in town who despised boomers, although this attitude may be attributed as much to their envy of boomers' big cars and money as to the fact that comparatively few boomers were teetotalers or celibates. On the other hand, it would be equally misleading to assume that there were not some boomers who were quiet, modest men—maybe as many as six or seven—one of them being, for instance, a big quiet Kentuckian named Ace Cowan (whose wife was with him in Michigan), and another being Johnny Atkins, who once at the Nicolet drank a dozen double Martinis without causing a fuss or seeming drunk, and then floated quietly, happily out into the night.

And there was also Jack Kelly, the tall 235-pound son of a Philadelphia sailmaker, who, despite years of work on noisy bridges and despite getting hit on the head by so much falling hardware that he had fifty-two stitches in his scalp, remained ever mild. And finally there was another admired man on the Mackinac—the superintendent, Art "Drag-Up" Drilling, a veteran boomer from Arkansas who went West to work on the Golden Gate and Oakland Bay bridges in the thirties, and who was called "Drag-Up" because he always said, though never in threat, that he'd sooner drag-up and leave town than work under a superintendent who knew less about bridges than he.

So he went from town to town, bridge to bridge, never really satisfied until he became the top bridgeman—as he did on the Mackinac, and as he hoped to do in 1962 on the Verrazano-Narrows Bridge.

In the course of his travels, however, Drag-Up Drilling sired a son named John. And while John Drilling inherited much of his father's soft Southern charm and easy manner, these qualities actually belied the devil beneath. For John Drilling, who was only nineteen years old when he first joined the gang on the Mackinac, worked as hard as any to leave the boomer's mark on St. Ignace.

John Drilling had been born in Oakland in 1937 while his father was finishing on the Bay Bridge there. And in the next nineteen years he followed his father, living in forty-one states, attending two dozen schools, charming the girls—marrying one, and living with her for four months. There was nothing raw nor rude in his manner. He was always extremely genteel and clean-cut in appearance, but, like many boomers' offspring, he was afflicted with what old bridgemen call "rambling fever."

This made him challenging to some women, and frustrating to others, yet intriguing to most. On his first week in St. Ignace, while stopped at a gas station, he noticed a carload of girls nearby and, exuding all the shy and bumbling uncertainty of a new boy in town, addressed himself politely to the prettiest girl in the car—a Swedish beauty, a very healthy girl whose boyfriend had just been drafted—and thus began an unforgettable romance that would last until the next one.

Having saved a few thousand dollars from working on the Mackinac, he became, very briefly, a student at the University of Arkansas and also bought a $2,700 Impala. One night in Ola, Arkansas, he cracked up the car and might have gotten into legal difficulty had not his date that evening been the judge's daughter.

John Drilling seemed to live a charmed life. Of all the bridge builders who worked on the Mackinac, and who would later come east to work on the Verrazano-Narrows Bridge, young John Drilling seemed the luckiest—with the possible exception of his close friend, Robert Anderson.

Anderson was luckier mainly because he had lived longer, done more, survived more; and he never lost his sunny disposition or incurable optimism. He was thirty-four years old when he came to the Mackinac. He had been married to one girl for a dozen years, to another for two weeks. He had been in auto accidents, been hit by falling tools, taken falls—once toppling forty-two feet—but his only visible injury was two missing inside fingers on his left hand, and he never lost its full use.

One day on the north tower of the Mackinac, the section of catwalk upon which Anderson was standing snapped loose, and suddenly it came sliding down like a rollercoaster, with Anderson clinging to it as it bumped and raced down the cables, down 1,800 feet all the way to near the bottom where the cables slope gently and straighten out before the anchorage. Anderson quietly got off and began the long climb up again. Fortunately for him, the Mackinac was designed by David B. Steinman, who preferred long, tapering backspans; had the bridge been designed by O. H. Ammann, who favored shorter, chunkier backspans, such as the type he was then creating for the Verrazano-Narrows Bridge, Bob Anderson would have had a steeper, more abrupt ride down, and might have gone smashing into the anchorage and been killed. But Anderson was lucky that way.

Off the bridge, Anderson had a boomer's luck with women. All the moving around he had done during his youth as a boomer's son, all the shifting from town to town and the enforced flexibility required of such living, gave him a casual air of detachment, an ability to be at home anywhere. Once, in Mexico, he made his home in a whorehouse. The prostitutes down there liked him very much, fought over him, admired his gentle manners and the fact that he treated them all like ladies. Finally the madam invited him in as a full-time house guest and each night Anderson would dine with them, and in the morning he stood in line with them awaiting his turn in the shower.

Though he stands six feet and is broad-shouldered and erect, Bob Anderson is not a particularly handsome fellow; but he has bright, alert eyes, and a round, friendly, usually smiling face, and he is very disarming, a sort of Tom Jones of the bridge business—smooth and swift, somewhat gallant, addicted to good times and hot-blooded women, and yet never slick or tricky.

He is also fairly lucky at gambling, having learned a bit back in Oklahoma from his uncle Manuel, a guitar-playing rogue who once won a whole carnival playing poker. Anderson avoids crap games, although one evening at the Nicolet, when a crap game got started on the floor of the men's room and he'd been invited to join, he did.

"Oh, I was drunk that night," he said, in his slow southwestern drawl, to a friend some days later. "I was so drunk I could hardly see. But I jes' kept rolling them dice, and all I was seeing was sevens and elevens, sevens and elevens, *Jee-sus Kee-rist*, all night long it went like that, and I kept winning and drinking and winning some more. Finally lots of other folks came jamming in, hearing all the noise and all, and in this men's toilet room there's some women and tourists who also came in jes' watching me roll those sevens and elevens.

"Next morning I woke up with a helluva hangover, but on my bureau I seen this pile of money. And when I felt inside my pockets they were stuffed with bills, crumpled up like dried leaves. And when I counted it all, it came to more than one thousand dollars. And that day on the bridge, there was guys coming up to me and saying, 'Here, Bob, here's the fifty I borrowed last night,' or, 'Here's the hundred,' and I didn't even remember they borrowed it. *Jee-sus Kee-rist*, what a night!"

When Bob Anderson finally left the Mackinac job and St. Ignace, he had managed to save five thousand dollars, and, not knowing what else to do with it, he bought a round-trip airplane ticket and went flying off to Tangier, Paris, and Switzerland—"whoring and drinking," as he put it—and then, flat broke, except for his return ticket, he went back to St. Ignace and married a lean, lovely brunette he'd been unable to forget.

And not long after that, he packed his things and his new wife, and along with dozens of other boomers—with John Drilling and Drag-Up, with Ace Cowan and Jack Kelly and other veterans of the Mackinac and the Nicolet—he began the long road trip eastward to try his luck in New York.

PANIC IN BROOKLYN

"*You sonamabitch!*" the old Italian shoemaker cried, standing in the doorway of the Brooklyn real estate office, glaring at the men who sat behind desks in the rear of the room. "You *sonamabitch*," he repeated when nobody looked up.

"*Hey,*" snapped one of the men, jumping up from his desk, "who are you talking to?"

"You," said the shoemaker, his small, disheveled figure leaning against the door unsteadily, as if he'd been drinking, his tiny dark eyes angry and bloodshot. "You take-a my store . . . you no give-a me notting, you . . ."

"Now listen here," said the real estate man, quickly walking to where the shoemaker stood and looking down at him hard, "we will have none of *that* talk around here. In fact I am going to call the cops . . ."

He grabbed the phone nearest him and began to dial. The shoemaker watched for a moment, not seeming to care. Then he shrugged to himself and slowly turned and, without another word, walked out the door and shuffled down the street. The real estate man, putting down the telephone, watched the shoemaker go. He did not chase him. He wanted nothing further to do with him— neither with him nor with *any* of those boisterous people who had been making so much noise lately, cursing or signing petitions or is- suing threats, as if it had been the *real estate men's* idea to build the Verrazano Narrows Bridge and the big highway leading up to it, the highway that would cut into the Bay Ridge section of Brooklyn where seven thousand people now lived, where eight hundred buildings now stood—including a shoestore—and would level everything in its path into a long, smooth piece of concrete.

No, it was not their idea, it was the idea of Robert Moses and his Triborough Bridge and Tunnel Authority to build the bridge and its adjoining highways—but the real estate men, hired by the Authority, were getting most of the direct blame because it was they, not Moses, who had to face the people and say, "Abandon your homes—we must build a bridge."

Some people, particularly old people, panicked. Many of them pleaded with the Authority's representatives and prayed to God not to destroy these homes where their children had been born, where their husbands had died. Others panicked with anger, saying this was their home, their castle, and Mr. Moses would have to drag them from it bodily.

Some took the news quietly, waiting without words to be

listed among the missing waiting for the moving van as if it meant death itself. With the money the Authority paid them for their old home, they went to Florida, or to Arizona, or to another home in Brooklyn, any home, not seeming to care very much because now they were old people and new homes were all the same.

The old shoemaker, nearly seventy, returned to Southern Italy, back to his native Cosenza, where he had some farmland he hoped to sell. He had left Cosenza for America when he was twenty-two years old. And now, in 1959, seeing Cosenza again was seeing how little it had changed. There were still goats and donkeys climbing up the narrow roads, and some peasant women carrying clay pots on their heads, and a few men wearing black bands on their sleeves or ribbons in their lapels to show that they were in mourning; and still the same white stone houses speckled against the lush green of the mountainside—houses of many generations.

When he arrived, he was greeted by relatives he had long forgotten, and they welcomed him like a returning hero. But later they began to tell him about their ailments, their poverty, all their problems, and he knew what was coming next. So he quickly began to tell them about his problems, sparing few details, recalling how he had fallen behind in the rent of his shoestore in Brooklyn, how the Authority had thrown him out without a dime, and how he now found himself back in Italy where he had started—all because this damned bridge was going to be built, this bridge the Americans were planning to name after an Italian explorer the shoemaker's relatives had never heard of, this Giovanni da Verrazano, who, sailing for the French in 1524, discovered New York Bay. The shoemaker went on and on, gesturing with his hands and making his point, making certain they knew he was no soft touch—and, a day or two later, he went about the business of trying to sell the farmland. . . .

On the Staten Island side, opposition to the bridge was nothing like it was in Brooklyn, where more than twice as many people and buildings were affected by the bridge; in fact, in Staten Island there had long been powerful factions that dreamed of the day when a bridge might be built to link their borough more firmly with the rest of New York City. Staten Island had always been the most isolated, the most ignored of New York's five boroughs; it was separated from Manhattan by five miles of water and a half-hour's ride on the ferry.

While New Yorkers and tourists had always enjoyed riding the Staten Island ferry—"a luxury cruise at a penny a mile"—nobody was ever much interested in getting to the other side. What was there to see? Sixty percent of the island's fifty-four square miles were underdeveloped as of 1958. Most of its 225,000 citizens lived in one-family houses. It was the dullest of New York's boroughs, and when a New York policeman was in the doghouse with headquarters, he was often transferred to Staten Island.

The island first acquired its rural quality when the British controlled it three hundred years ago, encouraging farming rather than manufacturing, and that was the way many Staten Islanders wanted it to remain—quiet and remote. But on the last day of 1958, after years of debate and doubt, plans for the building of the Verrazano-Narrows Bridge finally became definite and the way of those who cherished the traditional life was in decline. But many more Staten Island residents were overjoyed with the news; they had wanted a change, had grown bored with the provincialism, and now hoped the bridge would trigger a boom—and suddenly they had their wish.

The bridge announcement was followed by a land rush. Real estate values shot up. A small lot that cost $1,200 in 1958 was worth $6,000 in 1959, and larger pieces of property worth $100,000 in the morning often sold for $200,000 that afternoon. Tax-delinquent properties were quickly claimed by the city. Huge foreign syndicates from Brazil, Italy, and Switzerland moved in for a quick kill. New construction was planned for almost every part of Staten Island, and despite complaints and suits against contractors for cheaply built homes (one foreman was so ashamed of the shoddy work he was ordered to do that he waited until night to leave the construction site) nothing discouraged the boom or deglamorized the bridge in Staten Island.

The bridge had become, in early 1959, months before any workmen started to put it up, the symbol of hope.

"We are now on our way to surmounting the barrier of isolation," announced the borough president, Albert V. Maniscalco—while other leaders were conceding that the bridge, no matter what it might bring, could not really hurt Staten Island. What was there to hurt? "Nothing has ever been successful in Staten Island in its entire history," said one resident, Robert Regan, husband of opera singer Eileen Farrell. He pointed out that there had been attempts in the past to establish a Staten Island opera company, a semi-professional football team, a dog track, a boxing arena, a symphony orchestra, a midget auto track, a basketball team—and all failed. "The only thing that might save this island," he said, "is a lot of new people."

Over in Brooklyn, however, it was different. They did not need or want new people. They had a flourishing, middle-class, almost all-white community in the Bay Ridge section, and they were satisfied with what they had. Bay Ridge, which is in western Brooklyn along

the ridge of Upper New York Bay and Lower New York Bay, commands a superb view of the Narrows, a mile-wide tidal strait that
connects the two bays, and through which pass all the big ships en-
tering or leaving New York. Among its first settlers were thousands
of Scandinavians, most of them Danes, who liked Bay Ridge because of its nearness to the water and the balmy breeze. And in the
late nineteenth century, Bay Ridge became one of the most exclu-
sive sections of Brooklyn.

It was not that now, in 1959, except possibly along its shorefront section, which was lined with trees and manicured lawns and
with strong sturdy homes, one of them occupied by Charles Atlas.
The rest of Bay Ridge was almost like any other Brooklyn residential neighborhood, except that there were few if any Negroes living
among the whites. The whites were mostly Catholic. The big
churches, some with parishes in excess of 12,000, were supported
by the lace-curtain Irish and aspiring Italians, and the politics, usually Republican, were run by them, too. There were still large
numbers of Swedes and Danes, and also many Syrian shopkeepers,
and there were old Italian immigrants (friends of the shoemaker)
who were hanging on, but it was the younger, second- and third-
generation Italians, together with the Irish, who determined the
tone of Bay Ridge. They lived, those not yet rich enough for the
shorefront homes, in smaller brown brick houses jammed together
along tree-lined streets, and they competed each day for a parking
place at the curb. They shopped along busy sidewalks clustered
with tiny neighborhood stores with apartments above, and there
were plenty of small taverns on corners, and there was the
Hamilton House for a good dinner at night—provided they wore a
jacket and tie—and there was a dimly lit sidestreet supper-club on

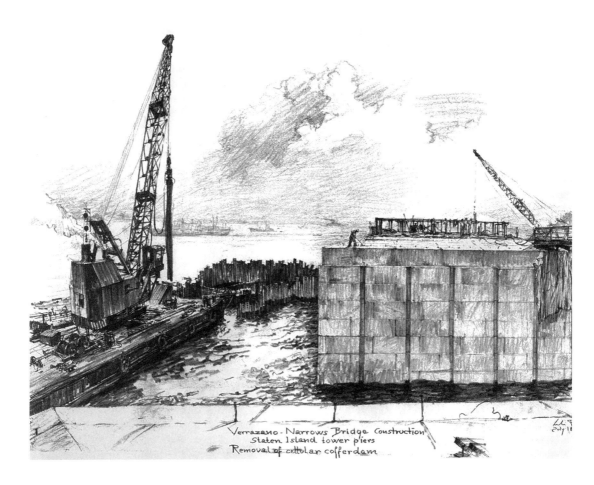

Verrazano-Narrows Bridge Construction
Staten Island tower piers
Removal of cellular cofferdam

the front barstool of which sat a curvesome, wrinkled platinum
blond with a cigarette, but no match.

So Bay Ridge, in 1959, had things in balance; it was no
longer chic, but it was tidy, and most people wanted no change, no
new people, no more traffic. And they certainly wanted no bridge.

When the news came that they would get one, the local politicians were stunned. Some women began to cry. A number of people refused to believe it. They had heard this talk before, they said, pointing out that as far back as 1888 there had been plans for a railroad tunnel that would link Brooklyn and Staten Island. And in 1923 New York's Mayor Hylan even broke ground for a combined rail-and-automobile tunnel to Staten Island, and all that happened was that the city lost a half-million dollars and now has a little hole somewhere going nowhere.

And there had been talk about this big bridge across the Narrows for *twenty years*, they said, and each time it turned out to be just talk. In 1950 there was talk that a bridge between Brooklyn and Staten Island was a good thing, but what if the Russians blew it up during a war: Would not the United States Navy ships docked in New York harbor be trapped behind the collapsed bridge at the harbor's entrance? And a year later, there was more talk of a tunnel to Staten Island, and then more debate on the bridge, and it went on this way, on and on. So, they said, in 1959, maybe this is *still* all talk, no action, so let's not worry.

What these people failed to realize was that about 1957 the talk changed a little; it became more intense, and Robert Moses was getting more determined, and New York City's Fire Commissioner was so sure in 1957 that the bridge to Staten Island would become a reality that he quickly got in his bid with the City Planning Commission for a big new Staten Island firehouse, asking that he be given $379,500 to build it and $250,000 to equip it. They did not realize that the powerful Brooklyn politician Joseph T. Sharkey saw the bridge as inevitable in 1958, and he had made one last desperate attack, too late, on Robert Moses on the City Council floor, shouting that Moses was getting too much power and was listening only to the engineers, not to the will of the people. And they did not realize, too, that while they were thinking it was still *all talk*, a group of engineers around a drawing board were quietly inking out a large chunk of Brooklyn that would be destroyed for the big approachway to the bridge—and one of the engineers, to his horror, realized that his plan included the demolition of the home of his own mother-in-law. When he told her the news, she screamed and cried and demanded he change the plan. He told her he was helpless to do so; the bridge was inevitable. She died without forgiving him.

The bridge was inevitable—and it was inevitable they would hate it. They saw the coming bridge not as a sign of progress, but as a symbol of destruction, as an enormous sea monster that soon would rise out of the water and destroy eight hundred buildings and force seven thousand Bay Ridge people to move—all sorts of people: housewives, bartenders, a tugboat skipper, doctors, lawyers, a pimp, teetotalers, drunks, secretaries, a retired light-heavyweight fighter, a former Ziegfeld Follies girl, a family of seventeen children (two dogs and a cat), a dentist who had just spent $13,000 installing new chairs, a vegetarian, a bank clerk, an assistant school principal, and two lovers, a divorced man of forty-one and an unhappily married woman who lived across the street. Each afternoon in his apartment they would meet, these lovers, and make love and wonder what next, wonder if she could ever tell her husband and leave her children. And now, suddenly, this bridge was coming between the lovers, would destroy their neighborhood and their quiet afternoons together, and they had no idea, in 1959, what they would do. What the others did, the angry ones, was join the "Save Bay Ridge" committee, which tried to fight Moses until the bulldozers were bashing down their doors. They signed petitions, and made speeches, and screamed, "This bridge—who needs it?" News photographers took their photographs and reporters interviewed them, quoting their impassioned pleas, and Robert Moses became furious.

He wrote letters to a newspaper publisher saying that the reporter had distorted the truth, had lied, had emphasized only the bad part, not the good part, of destroying people's homes. Most people in Brooklyn did not, in 1959, understand the good part, and so they held on to their homes with determination. But sooner or later, within the next year or so, they let go. One by one they went, and soon the house lights went out for the last time, and then moving vans rolled in, and then the bulldozers came crashing up and the walls crumbled down, and the roofs caved in and everything was hidden in an avalanche of dust—a sordid scene to be witnessed by the hold-out next door, and soon he, too, would move out, and then another, and another. And that is how it went on each block, in each neighborhood, until, finally, even the most determined hold-out gave in because, when a block is almost completely destroyed, and one is all alone amid the chaos, strange and unfamiliar fears sprout up: the fear of being alone in a neighborhood that is dying; the fear of a band of young vagrants who occasionally would roam through the rubble smashing windows or stealing doors, or picket fences, light fixtures, or shrubbery, or picking at broken pictures or leftover love letters; fear of the derelicts who would sleep in the shells of empty apartments or hanging halls; fear of the rats that people said would soon be crawling up from the shattered sinks or sewers because, it was explained, rats also were being dispossessed in Bay Ridge, Brooklyn.

One of the last hold-outs was a hazel-eyed, very pretty brunette divorced woman of forty-two named Florence Campbell. She left after the lovers, after the dentist, and after the former Ziegfeld Follies girl, Bessie Gros Dempsey, who had to pack up her

350 plumed hats and old scrapbooks; she left after the crazy little man who had been discovered alone in an empty apartment house because, somehow, he never heard the bulldozers beneath him and had no idea that a bridge was being built.

She left after the retired prizefighter, Freddy Fredericksen, who had only lost twice before, and after Mr. and Mrs. John G. Herbert, the parents of seventeen children—although Florence Campbell's leaving was nowhere as complex as the Herberts'. It had taken them twelve trips to move all their furniture, all the bicycles, sleds, dishes, dogs to their new house a little more than a mile away—twelve trips and sixteen hours; and when they had finally gotten everything there, Mr. Herbert, a Navy Yard worker, discovered that the cat was missing. So early the next morning he sent two sons back to the old house, and they discovered the cat beneath the porch. They also discovered an old axe there. And for the next hour they used the axe to destroy everything they could of the old house; they smashed windows, walls, the floors, they smashed their old bedroom, the kitchen shelves, and the banister of the porch, where they used to gather on summer nights, and they smashed without knowing exactly why, only knowing, as they took turns swinging, that they felt a little wild and gleeful and sad and mad as they smashed—and then, too tired to continue, they retrieved the cat from under the smashed porch and they left their old home for the last time.

In the case of Florence Campbell, it took more than even a murder to make her abandon her home. She had been living, since her divorce, with her young son in a sixty-four-dollar-a-month spacious apartment. It was difficult for her to find anything like it at a rental she could afford. The relocation agent, who had lost patience with her for turning down apartments he considered suitable but she considered too expensive, now forgot about her, and she was on her own to search alone at night after she had returned from her bookkeeping job with the Whitehall Club in Manhattan.

Then one morning she started smelling a strange odor in the apartment. She thought perhaps that her son had gone fishing the day before, after school, and had dumped his catch in the dumbwaiter. He denied it, and the next night, when the odor became worse, she telephoned the police. They soon discovered that the elderly man living on the first floor, the only other tenant in the house, had three days before murdered his wife with shotgun bullets and now, dazed and silent, he was sitting next to the corpse, empty whiskey bottles at his feet.

"Lady, do me a favor," the police sergeant said to Florence Campbell. "Get out of this block, will ya?"

She said she would, but she still could not find an apartment during her searchings. She had no relatives she could move in with, no friends within the neighborhood, because they had all moved. When she came home at midnight from apartment hunting, she would find the hall dark—somebody was always stealing the light bulb—or she might stumble over a drunken derelict sleeping on the sidewalk in front of the downstairs door.

A few nights after the sergeant's warning she was awakened from sleep by the sounds of shuffling feet outside her door and the pounding of fists against the wall. Her son, in the adjoining bedroom, jumped up, grabbed a shotgun he kept in his closet, and ran out into the hall. But it was completely dark, the light bulb had been stolen again. He tripped and Florence Campbell screamed.

A strange man raced up the steps to the roof. She called the police. They came quickly but could find no one on the roof. The police sergeant again told her to leave, and she nodded, weeping, that she would. The next day she was too nervous to go to work, and so she went to a nearby bar to get a drink and told the bartender what had happened, and, very excited, he told her he knew of an apartment that was available a block away for sixty-eight dollars a month. She ran to the address, got the apartment—and the land-lord could not understand why, after she got it, she began to cry.

SURVIVAL OF THE FITTEST

The bridge began as bridges always begin—silently. It began with underwater investigations and soil studies and survey sheets; and when the noise finally started, on January 16, 1959, nobody in Brooklyn or Staten Island heard it.

It started with the sound of a steam pile driver ramming a pipe thirty-six inches in diameter into the silt of a small island off the Brooklyn shore. The island held an old battered bastion called Fort Lafayette, which had been a prison during the Civil War, but now it was about to be demolished, and the island would only serve as a base for one of the bridge's two gigantic towers.

Nobody heard the first sounds of the bridge because they were soft and because the island was six hundred feet off the

Brooklyn shore; but even if it had been closer, the sounds would not have risen above the rancor and clamor of the people, for when the drilling began, the people still were protesting, still were hopeful that the bridge would never be built. They were aware that the city had not yet formally condemned their property—but that came three months later. On April 30, 1959, in Brooklyn Supreme Court, Justice J. Vincent Keogh—who would later go to jail on charges of sharing in a bribe to fix another case—signed the acquisition papers, and four hundred Bay Ridge residents suddenly stopped protesting and submitted in silence.

The next new noise was the spirited, high-stepping sound of a marching band and the blaring platitudes of politicians echoing over a sun-baked parade ground on August 14, 1959—it was ground-breaking day for the bridge, with the ceremony held, wisely, on the Staten Island side. Over in Brooklyn, when a reporter asked State Senator William T. Conklin for a reaction, the Bay Ridge representative snapped, "It is not a ground-breaking—to many it will be heart-breaking." And then, slowly and more emotionally, he continued: "Any public official attending should always be identified in the future with the cruelty that has been inflicted on the community in the name of progress."

Governor Nelson Rockefeller of New York had been invited to attend the ceremony in Staten Island, but he sent a telegram expressing regret that a prior engagement made it impossible for him to be there. He designated Assembly Speaker Joseph Carlino to read his message. But Mr. Carlino did not show up. Robert Moses had to read it.

As Mr. Moses expressed all the grand hope of the future, a small airplane chartered by the Staten Island Chamber of Commerce circled overhead with an advertising banner that urged

"Name it the Staten Island Bridge." Many people opposed the name Verrazano—which had been loudly recommended by the Italian Historical Society of America and its founder, John N. La Corte—because they could not spell it. Others, many of them Irish, did not want a bridge named after an Italian, and they took to calling it the "Guinea Gangplank." Still others advocated simpler names—"The Gateway Bridge," "Freedom Bridge," "Neptune Bridge," "New World Bridge," "The Narrows Bridge." One of the last things ever written by Ludwig Bemelmans was a letter to the *New York Times* expressing the hope that the name "Verrazano" be dropped in favor of a more "romantic" and "tremendous" name, and he suggested calling it the "Commissioner Moses Bridge." But the Italian Historical Society, boasting a large membership of emotional voters, was not about to knuckle under, and finally after months of debate and threats, a compromise was reached in the name "Verrazano-Narrows Bridge."

The person making the least amount of noise about the bridge all this time was the man who was creating it—Othmar H. Ammann, a lean, elderly, proper man in a high starched collar, who now, in his eightieth year, was recognized as probably the greatest bridge engineer in the world. His monumental achievement so far, the one that soared above dozens of others, was the George Washington Bridge, the sight of which had quietly thrilled him since its completion in 1931. Since then, when he and his wife drove down along the Hudson River from upstate New York and suddenly saw the bridge looming in the distance, stretching like a silver rainbow over the river between New York and New Jersey, they often gently bowed and saluted it.

"That bridge is his firstborn, and it was a difficult birth," his wife once explained. "He'll always love it best." And Othmar Ammann, though reluctant to reveal any sentimentality, nevertheless once described its effect upon him. "It is as if you have a beautiful daughter," he said, "and you are the father."

But now the Verrazano-Narrows Bridge presented Ammann with an even larger task. And to master its gigantic design he would even have to take into account the curvature of the earth. The two 693-foot towers, though exactly perpendicular to the earth's surface, would have to be one and five-eighths inches farther apart at their summits than at their bases.

Though the Verrazano-Narrows Bridge would require 188,000 tons of steel—three times the amount used in the Empire State Building—Ammann knew that it would be an ever restless structure, would always sway slightly in the wind. Its steel cables would swell when hot and contract when cold, and its roadway would be twelve feet closer to the water in summer than in winter. Sometimes, on long hot summer days, the sun would beat down on one side of the structure with such intensity that it might warp the steel slightly, making the bridge a fraction lower on its hot side than on its shady side. So, Ammann knew, any precision measuring to be done during the bridge's construction would have to be done at night.

From the start of a career that began in 1902, when he graduated from the Swiss Federal Polytechnic Institute with a degree in civil engineering, Ammann had made few mistakes. He had been a careful student, a perfectionist. He had witnessed the rise and fall of other men's creations, had seen how one flaw in mathematics could ruin an engineer's reputation for life—and he was determined it would not happen to him.

Othmar Hermann Ammann had been born on March 26, 1879, in Schaffhausen, Switzerland, into a family that had been established in Schaffhausen since the twelfth century. His father had been a prominent manufacturer and his forebears had been physicians, clergymen, lawyers, government leaders, but none had been engineers, and few had shared his enthusiasm for bridges.

There had always been a wooden bridge stretching from the village of Schaffhausen across the Rhine, the most famous of them being built at a length of 364 feet in the 1700s by a Swiss named Hans Ulrich Grubenmann. It had been destroyed by the French in 1799, but had been replaced by others, and as a boy Othmar Ammann saw bridges as a symbol of challenge and a monument to beauty.

In 1904, after working for a time in Germany as a design engineer, Ammann came to the United States—which, after slumbering for many decades in a kind of dark age of bridge design, was now finally experiencing a renaissance. American bridges were getting bigger and safer; American engineers were now bolder than any in the world.

There were still disasters, but it was nothing like it had been in the middle 1800s, when as many as forty bridges might collapse in a single year, a figure that meant that for every four bridges put up one would fall down. Usually it was a case of engineers not knowing precisely the stress and strain a bridge could withstand, and also there were cases of contractors being too cost-conscious and willing to use inferior building materials. Many bridges in those days, even some railroad bridges, were made of timber. Others were made of a new material, wrought iron, and nobody knew exactly how it would hold up until two disasters—one in Ohio, the other in Scotland—proved its weakness.

The first occurred on a snowy December night in 1877 when a train from New York going west over the Ashtabula Bridge in Ohio suddenly crumbled the bridge's iron beams and then, one by one, the rail cars fell into the icy waters, killing ninety-two people. Two years later, the Firth of Tay Bridge in Scotland collapsed under the strain of a locomotive pulling six coaches and a brakeman's van. It had been a windy Sunday night, and seventy-five people were killed, and religious extremists blamed the railroad for running trains on Sunday. But engineers realized that it was the wrought iron that was wrong, and these two bridge failures hastened the acceptance of steel—which has a working strength twenty-five percent greater than wrought iron—and thus began the great era that would influence young Othmar Ammann.

This era drew its confidence from two spectacular events—the completion in 1874 of the world's first steel bridge, a triple arch over the Mississippi River at St. Louis designed and built by James Buchanan Eads; and the completion in 1883 of the Brooklyn Bridge, first steel cable suspension span, designed by John Roebling and, upon his tragic death, completed by his son, Washington Roebling. Both structures would shape the future course in American bridge-building, and would establish a foundation of knowledge, a link of trial and error, that would guide every engineer through the twentieth century. The Roeblings and James Buchanan Eads were America's first heroes in high steel.

James B. Eads was a flamboyant and cocky Indiana boy whose first engineering work was raising sunken steamers from the bottom of the Mississippi. He also was among the first to explore the river's bed in a diver's suit, and he realized, when it came time for him to start constructing the foundations for his St. Louis bridge, that he could not rely on the Mississippi River soil for firmness, because it had a peculiar and powerful shifting movement.

So he introduced to America the European pneumatic caisson—an airtight enclosure that would allow men to work underwater without being hindered by the shifting tides. Eventually, as the caisson sank deeper and deeper and the men dug up more and more of the riverbed below, the bridge's foundation could penetrate the soft sand and silt and could settle solidly on the hard rock beneath the Mississippi. Part of this delicate operation was helped by Eads's invention—a sand pump that could lift and eject gravel, silt, and sand from the caisson's chamber.

Before Eads's bridge would be finished, however, 352 workmen would suffer from a strange new ailment—caisson's disease or "the bends"—and twelve men would die from it, and two would remain crippled for life. But from the experience and observations made by James Eads's physician, Dr. Jaminet, who spent time in the caisson with the men and became temporarily paralyzed himself, sufficient knowledge was obtained to greatly reduce the occurrence of the ailment on future jobs.

When the St. Louis steel bridge was finished, James Eads, to show its strength, ran fourteen locomotives across each of the bridge's three arches. Later a fifteen-mile parade marched across it, President Grant applauded from the reviewing stand, General Sherman drove in the last spike on the Illinois side, and Andrew Carnegie, who had been selling bonds for the project, made his first fortune. The bridge was suddenly instrumental in the development of St. Louis as the most important city on the Mississippi River, and it helped develop the transcontinental railroad systems. It was credited with "the winning of the West" and was pictured on a United States stamp in 1898; and in 1920 James Buchanan Eads became the first engineer elected to the American Hall of Fame.

He died an unhappy man. A project he envisioned across the Isthmus of Tehuantepec did not work out.

John Augustus Roebling was a studious German youth born in 1806, in a small town called Muhlhausen, to a tobacco merchant who smoked more than he sold and to a mother who prayed he would someday amount to more than his father. Largely through her ambition and thrift he received a fine education in architecture and engineering in Berlin, and later he worked for the Prussian government building roads and bridges.

But there was little opportunity for originality, and so at the age of twenty-five he came to America and soon, in Pennsylvania, he was working as a surveyor for the railroads and canals. And one day, while observing how the hemp rope that hauled canal boats often broke, John Roebling began to experiment with a more durable fiber, and soon he was twisting iron wire into the hemp—an idea that would eventually lead him and his family into a prosperous industry that today, in Trenton, New Jersey, is the basis for the Roebling Company—world's largest manufacturer of wire rope and cable.

But in those days it led John Roebling toward his more immediate goal, the construction of suspension bridges. He had seen smaller suspensions, hung with iron chains, during his student days in Germany, and he wondered if the suspension bridge might not be more graceful, longer, and stronger with iron wire rope, maybe even strong enough to support rail cars.

He had his chance to find out when, in 1851, he received a commission to build a suspension bridge over Niagara Falls. This opportunity arose only because the original engineer had abandoned the project after a financial dispute with the bridge company—this engineer being a brilliant but wholly unpredictable and

daring man named Charles Ellet. Ellet, when confronted with the problem of getting the first rope across Niagara, found the solution by offering five dollars to any boy who could fly a kite across it. Ellet later had a basket carrier made and he pulled himself over the rushing waters of Niagara to the other side; and next he did the same thing accompanied by his horse, as crowds standing on the cliffs screamed and some women fainted.

Things quieted down when Ellet left Niagara, but John Roebling, in his methodical way, got the job done. "Engineering," as

Joseph Gies, an editor and bridge historian, wrote, "is the art of the efficient, and the success of an engineering project often may be measured by the absence of any dramatic history." In 1855, Roebling's 821-foot single span was finished, and on March 6 of that year a 368-ton train crossed it—the first train in history to cross a span sustained by wire cables. The success quickly led Roebling to other bridge commissions, and in 1867 he started his greatest task, the Brooklyn Bridge.

It would take thirteen years to complete the Brooklyn Bridge, and both John Roebling and his son would be its victims. One summer morning in 1869, while standing on a pier off Manhattan, surveying the location of one of the towers, and paying no attention to the docking ferryboat that was about to bump into the pier, John Roebling suddenly had his foot caught and crushed between the pier floor and piles; tetanus set in, and two weeks later, at the age of sixty-three, he died.

At the death of his father, Washington Roebling, then thirty-two years old and the chief engineering assistant for the bridge, took over the job. Roebling had previously supervised the construction of other bridges that his father had designed, and had served as an engineering officer for the Union Army during the Civil War. During the war he had also been one of General Grant's airborne spies, ascending in a balloon to watch the movement of Lee's army during its invasion of Pennsylvania.

When he took over the building of the Brooklyn Bridge, Washington Roebling decided that since the bridge's tower foundations would have to be sunk forty-four feet into the East River on the Brooklyn side and seventy-six feet on the New York side, he would use pneumatic caissons—as James Eads had done a few years before with his bridge over the Mississippi. Roebling drove himself relentlessly, working in the caissons day and night and he finally collapsed. When he was carried up, he was paralyzed for life. He was then thirty-five years old.

But Washington Roebling, assisted by his wife, Emily, continued to direct the building of the bridge from his sickbed; he would watch the construction through field glasses while sitting at the window of his home on the Brooklyn shore; and then his wife—to whom he had taught the engineer's language, and who understood the problems involved—would carry his instructions to the superintendents on the bridge itself.

Washington Roebling was the first bridge engineer to use steel wire for his cables—it was lighter and stronger than the iron wire cables used by his father on the Niagara bridge—and he had every one of the 5,180 wires galvanized as a safeguard against rust. The first wire was drawn across the East River in 1877, and for the next twenty-six months, from one end of the bridge to the other, the small traveling wheels—looking like bicycle wheels with the tires missing—spun back and forth on pulleys, crossing the East River 10,360 times, each time bringing with them a double strand of wire which, when wrapped, would form the four cables that would hold up the center span of 1,595 feet and its two side spans of 930 feet each. This technique of spinning wire, and the use of a cowbell attached to each wheel to warn the men of its approaches, is still used today; it was used, in a more modern form, even by O. H. Ammann in the cable-spinning phase of his Verrazano-Narrows Bridge in the 1960s.

The Brooklyn Bridge was opened on May 24, 1883. Washington Roebling and his wife watched the celebrations from their windows through field glasses. It was a great day in New York—business was suspended, homes were draped with bunting, church bells rang out, steamships whistled. There was the thunder of guns from the forts in the harbor and from the Navy ships docked near the bridge, and finally, in open carriages, the dignitaries arrived. President Chester A. Arthur, New York's Governor Grover Cleveland, and the mayors of every city within several miles of New York arrived at the bridge. Later that night there was a procession in Brooklyn that led to Roebling's home, and he was congratulated in person by President Arthur.

To this day the Brooklyn Bridge has remained the most famous in America, and, until the Williamsburg Bridge was completed over the East River between Brooklyn and Manhattan in 1903, it was the world's longest suspension. In the great bridge boom of the twentieth century nineteen other suspension spans would surpass it—but none would cast a longer shadow. It has been praised by poets, admired by aesthetes, and sought by the suicidal. Its tower over the tenement roofs of the Lower East Side so electrified a young neighborhood boy named David Steinman that he became determined to emulate the Roeblings, and later he would become one of the world's great bridge designers; he alone, until his death in 1960, would challenge Ammann's dominance.

David Steinman at the age of fourteen had secured a pass from New York's Commissioner of Bridges to climb around the catwalks of the Williamsburg Bridge, then under construction, and he talked to bridge builders, took notes, and dreamed of the bridges he would someday build. In 1906, after graduating from City College in New York with the highest honors, he continued his engineering studies at Columbia, where, in 1911, he received his doctor's degree for his thesis on long-span bridges and foundations. Later he became consulting engineer on the design and construction of the Florianopolis Bridge in Brazil, the Mount Hope Bridge in Rhode Island, the

Grand Mere in Quebec, the Henry Hudson arch bridge in New York. It was Dr. David Steinman who was called upon to renovate the Brooklyn Bridge in 1948, and it was he who was selected over Ammann to build the Mackinac Bridge—although it was Ammann who emerged with the Verrazano-Narrows commission, the bridge that Steinman had dreamed of building.

The two men were never close as friends, possibly because they were too close in other ways. Both had been assistants in their earlier days to the late Gustav Lindenthal, designer of the Hell Gate and the Queensboro bridges in New York, and the two men were inevitably compared. They shared ambition and vanity, and yet possessed dissimilar personalities. Steinman was a colorfully blunt product of New York, a man who relished publicity and controversy, and who wrote poetry and had published books. Ammann was a stiff, formal Swiss gentleman, well born and distant. But that they were to be lifelong competitors was inevitable, for the bridge business thrives on competition; it exists on every level. There is competition between steel corporations as they bid for each job, and there is competition between even the lowliest apprentices in the work gangs. All the gangs—the riveters, the steel connectors, the cable spinners— battle throughout the construction of every bridge to see who can do the most work, and later in bars there is competition to see who can drink the most, brag the most. But here, on the lower level, among the bridge workers, the rivalry is clear and open; on the higher level, among the engineers, it is more secret and subtle.

Some engineers quietly go through life envying one another, some quietly prey on others' failures. Every time there is a bridge disaster, engineers who are unaffiliated with its construction flock to the site of the bridge and try to determine the reason for the failure. Then, quietly, they return to their own plans, armed with the knowledge of the disaster, and patch up their own bridges, hoping to prevent the same thing. This is as it should be. But it does not belie the truth of the competition. When a bridge fails, the engineer who designed it is as good as dead. In the bridge business, on every level, there is an endless battle to stay alive—and no one has stayed alive longer than O. H. Ammann.

Ammann was among the engineers who, in 1907, investigated the collapse of a cantilever bridge over the St. Lawrence River near Quebec. Eighty-six workmen, many of them Indians, who were just learning the high-construction business then, fell with the bridge, and seventy-five drowned. The engineer whose career ended with his failure was Theodore Cooper, one of America's most noted engineers—the same man who had been so lucky years before when, after falling one hundred feet into the Mississippi River while working on James Eads' bridge, not only survived but went back to work the same day.

But now, in 1907, it was the opinion of most engineers that Theodore Cooper did not know enough about the stresses involved in the cantilever bridge. None of them did. There is no way to know enough about bridge failure until enough bridges have failed. "This bridge failed because it was not strong enough," one engineer, C. C. Schneider, quipped to the others. Then they all returned to their own bridges, or to their plans for bridges, to see if they too had made miscalculations.

One bridge that perhaps was saved in this manner was Gustav Lindenthal's Queensboro Bridge, which was then approaching completion over the East River in Manhattan. After a re-examination, it was concluded that the Queensboro was inadequate to safely carry its intended load. So the four rapid-transit tracks that had been planned for the upper deck were reduced by two. The loss of the two tracks was compensated by the construction of a subway tunnel a block away from the bridge—the BMT tunnel at Sixtieth Street under the East River, built at an additional cost of $4,000,000.

In November of 1940, when the Tacoma Narrows Bridge fell into the waters of Puget Sound in the state of Washington, O. H. Ammann was again one of the engineers called in to help determine the cause. The engineer who caught the blame in this case was L. S. Moisseiff, a man with a fine reputation throughout the United States.

Moisseiff had been involved in the design of the Manhattan Bridge in New York, and had been the consulting engineer of the Ambassador Bridge in Detroit and the Golden Gate in California, among many others, and nobody had questioned him when he planned a lean, two-lane bridge that would stretch 2,800 feet over the waters of Puget Sound. True, it was a startlingly slim, fragile-looking bridge, but during this time there had been an aesthetic trend toward slimmer, sleeker, lovelier suspension bridges. This was the same trend that led David Steinman to paint his Mount Hope Bridge over Narragansett Bay a soft green color, and to have its cables strung with lights and approaches lined with evergreens and roses, costing an additional $70,000 for landscaping.

There was also a prewar trend toward economizing on the over-all cost of bridge construction, however, and one way to save money without spoiling the aesthetics—and supposedly without diminishing safety—was to shape the span and roadway floor with solid plate girders, not trusses that wind could easily pass through. And it was partially because of these solid girders that, on days when the wind beat hard against its solid mass of roadway, the Tacoma Narrows Bridge kicked up and down. But it never kicked too much, and the motorists, far from becoming alarmed, actually loved it, enjoyed riding over it. They knew that all bridges swayed a little in the wind—this bridge was just livelier, that was all, and they began calling it, affectionately, "Galloping Gertie."

Four months after it had opened—on November 7—with the wind between thirty-five and forty-two miles an hour, the bridge suddenly began to kick more than usual. Sometimes it would heave up and down as much as three feet. Bridge authorities decided to close the bridge to traffic; it was a wise decision, for later it began to twist wildly, rising on one side of its span, falling on the other, rising and falling sometimes as much as twenty-eight feet, tilting at a forty-five-degree angle in the wind. Finally, at 11 A.M., the main span ripped away from its suspenders and went crashing into Puget Sound.

The factors that led to the failure, the examining engineers deduced, were generally that the tall skinny bridge was too flexible and lacked the necessary stiffening girders; and also they spoke about a new factor that they had previously known very little about: "aerodynamic instability."

And soon, on other bridges, on bridges all over America and elsewhere, adjustments were made to compensate for the instability. The Golden Gate underwent alterations that cost more than $3,000,000. The very flexible Bronx-Whitestone Bridge in New York, which Leon Moisseiff had designed—with O. H. Ammann directing the planning and construction—had holes punched into its plate girders and had trusses added. Several other bridges that formerly had been slim and frail now became sturdier with trusses, and twenty years later, when Ammann was creating the Verrazano-Narrows Bridge, the Tacoma lesson lived on. Though the lower second deck on the Verrazano-Narrows was not yet needed, because the anticipated traffic could easily be accommodated by the six-lane upper deck, Ammann made plans for the second deck to go on right away—something he hadn't done in 1930 with his George Washington Bridge. The six-lane lower deck of the Verrazano will probably be without an automobile passenger for the next ten years, but the big bridge will be more rigid from its opening day.

After the Tacoma incident, Moisseiff's talents were no longer in demand. He never tried to pass off any of the blame on other engineers or the financiers; he accepted his decline quietly, though finding little solace in the fact that with his demise as an influential designer of bridges the world of engineering knowledge was expanded and bigger bridges were planned, bringing renown to others.

And so some engineers, like Leon Moisseiff and Theodore Cooper, go down with their bridges. Others, like Ammann and Steinman, remain high and mighty. But O. H. Ammann is not fooled by his fate.

One day, after he had completed his design on the Verrazano-Narrows Bridge, he mused aloud in his New York apartment, on the thirty-second floor of the Hotel Carlyle, that one reason he has experienced no tragedy with his bridges is that he has been blessed with good fortune.

"I have been lucky," he said, quietly.

"Lucky!" snapped his wife, who attributes his success solely to his superior mind.

"Lucky," he repeated, silencing her with his soft, hard tone of authority.

CHAPTER FOUR

PUNKS AND PUSHERS

Building a bridge is like combat; the language is of the barracks, and the men are organized along the lines of the noncommissioned officers' caste. At the very bottom, comparable to the Army recruit, are the apprentices—called "punks." They climb catwalks with buckets of bolts, learn through observation and turns on the tools, occasionally are sent down for coffee and water, seldom hear thanks. Within two or three years, most punks have become full-fledged bridgemen, qualified to heat, catch, or drive rivets; to raise, weld, or connect steel—but it is the last job, connecting the steel, that most captures their fancy. The steel connectors stand highest on the bridge, their sweat taking minutes to hit the ground, and when the derricks hoist up new steel, the connectors reach out and grab it with their hands, swing it into position, bang it with bolts and mallets, link it temporarily to the steel already in place, and leave the rest to the riveting gangs.

Connecting steel is the closest thing to aerial art, except the men must build a new sky stage for each show, and that is what makes it so dangerous—that and the fact that young connectors sometimes like to grandstand a bit, like to show the old men how it is done, and so they sometimes swing on the cables too much, or stand on unconnected steel, or run across narrow beams on windy days instead of straddling as they should—and sometimes they get so daring they die.

Once the steel is in place, riveting gangs move in to make it permanent. The fast, four-man riveting gangs are wondrous to watch. They toss rivets around as gracefully as infielders, driving in more than a thousand a day, each man knowing the others' moves, some having traveled together for years as a team. One man is called the "heater," and he sweats on the bridge all day over a kind of barbecue pit of flaming coal, cooking rivets until they are red— but not so red that they will buckle or blister. The heater must be a good cook, a chef, must think he is cooking sausages not rivets, because the other three men in the riveting gang are very particular people.

Once the rivet is red, but not too red, the heater tong-tosses it fifty, or sixty, or seventy feet, a perfect strike to the "catcher," who snares it out of the air with his metal mitt. Then the catcher relays the rivet to the third man, who stands nearby and is called the "busker-up"—and who, with a long cylindrical tool named after the anatomical pride of a stud horse, bucks the rivet into the prescribed hole and holds it there while the fourth man, the riveter, moves in from the other side and begins to rattle his gun against the rivet's front end until the soft tip of the rivet has been flattened and made round and full against the hole. When the rivet cools, it is as permanent as the bridge itself.

Each gang—whether it be a riveting gang, connecting gang or raising gang—is under the direct supervision of a foreman called a "pusher." (One night in a Brooklyn bar, an Indian pusher named Mike Tarbell was arrested by two plainclothes men who had overheard his occupation, and Tarbell was to spend three days in court and lose $175 in wages before convincing the judge that he was not a pusher of dope, only of bridgemen.)

The pusher, like an Army corporal who is bucking for sergeant, drives his gang to be the best, the fastest, because he knows that along the bridge other pushers are doing the same thing. They all know that the bridge company officials keep daily records of the productivity of each gang. The officials know which gang lifted the most steel, drove the most rivets, spun the most cable—and if the pusher is ambitious, wants to be promoted someday to a better job on the bridge, pushing is the only way.

But if he pushes too hard, resulting in accidents or death, then he is in trouble with the bridge company. While the bridge company encourages competition between gangs, because it wants to see the bridge finished fast, wants to see traffic jams up there and hear the clink of coins at toll gates, it does not want any accidents or deaths to upset the schedule or get into the newspapers or degrade the company's safety record with the insurance men. So the pusher is caught in the middle. If he is not lucky, if there is death in his gang, he may be blamed and be dropped back into the gang himself, and another workman will be promoted to pusher. But if he is lucky, and his gang works fast and well, then he someday might become an assistant superintendent on the bridge—a "walkin' boss."

The walkin' boss, of which there usually are four on a big bridge where four hundred or five hundred men are employed, commands a section of the span. One walkin' boss may be in charge of the section between an anchorage and a tower, another from that tower to the center of the span, a third from the center of the span to the other tower, the fourth from that tower to the other anchorage—and all they do all day is walk up and down, up and down, strutting like gamecocks, a look of suspicion in their eyes as they glance sideways to see that the pushers are pushing, the punks are punking, and the young steel connectors are not behaving like acrobats on the cables.

The thing that concerns walkin' bosses most is that they impress the boss, who is the superintendent, and is comparable to a top sergeant. The superintendent is usually the toughest, loudest, foulest-mouthed, best bridgeman on the whole job, and he lets everybody know it. He usually spends most of his day at a headquarters shack built along the shore near the anchorage of the bridge, there to communicate with the engineers, designers, and other white-collar officers from the bridge company. The walkin' bosses up on the bridge represent him and keep him informed, but about two or three times a day the superintendent will leave his shack and visit the bridge, and when he struts across the span the whole thing seems to stiffen. The men are all heads down at work, the punks seem petrified.

The superintendent selected to supervise the construction of the span and the building of the cables for the Verrazano-Narrows Bridge was a six-foot, fifty-nine-year-old, hot-tempered man named John Murphy, who, behind his back, was known as "Hard Nose" or "Short Fuse."

He was a broad-shouldered and chesty man with a thin strong nose and jaw, with pale blue eyes and thinning white hair—but the most distinguishing thing about him was his red face, a face so red that if he ever blushed, which he rarely did, nobody would know it. The red hard face—the result of forty years' booming in the high wind and hot sun of a hundred bridges and skyscrapers around America—gave Murphy the appearance of always being boiling mad at something, which he usually was.

He had been born, like so many boomers, in a small town without horizons—in this case, Rexton, a hamlet of three hundred in New Brunswick, Canada. The flu epidemic that had swept through Rexton in the spring of 1919, when Murphy was sixteen years old, killed his mother and father, an uncle and two cousins, and left him largely responsible for the support of his five younger brothers and sisters. So he went to work driving timber in Maine, and, when that got slow, he moved down to Pennsylvania and learned the bridge business, distinguishing himself as a steel connector because he was young and fearless. He was considered one of the best connectors on the George Washington Bridge, which he worked on in 1930 and 1931, and since then he had gone from one job to another, booming all the way up to Alaska to put a bridge across the Tanana River, and then back east again on other bridges and buildings.

In 1959 he was the superintendent in charge of putting up the Pan Am, the fifty-nine-story skyscraper in mid-Manhattan, and after that he was appointed to head the Verrazano job by the American Bridge Company, a division of United States Steel that had the contract to put up the bridge's span and steel cables.

When Hard Nose Murphy arrived at the bridge site in the

early spring of 1962, the long, undramatic, sloppy, yet so vital part of bridge construction—the foundations—was finished, and the two 693-foot towers were rising.

The foundation construction for the two towers, done by J. Rich Steers, Inc., and the Frederick Snare Corporation, if not an aesthetic operation that would appeal to the adventurers in high steel, nevertheless was a most difficult and challenging task, because the two caissons sunk in the Narrows had been among the largest ever built. They were 229 feet long and 129 feet wide, and each had sixty-six circular dredging holes—each hole being seventeen feet in diameter—and, from a distance, the concrete caissons looked like gigantic chunks of Swiss cheese.

Building the caisson that would support the pedestal which would in turn bear the foundation for the Staten Island tower had required 47,000 cubic yards of concrete, and before it settled on firm sand 105 feet below the surface, 81,500 cubic yards of muck and sand had to be lifted up through the dredging holes by clamshell buckets suspended from cranes. The caisson for the Brooklyn tower had to be sunk to about 170 feet below sea level, had required 83,000 cubic yards of concrete, and 143,600 cubic yards of muck and sand had to be dredged up.

The foundations, the ones that anchor the bridge to Staten Island and Brooklyn, were concrete blocks the height of a ten-story building, each triangular-shaped, and holding, within their hollows, all the ends of the cable strands that stretch across the bridge. These two anchorages, built by The Arthur A. Johnson Corporation and Peter Kiewit Sons' Company, hold back the 240,000,000-pound pull of the bridge's four cables.

It had taken a little more than two years to complete the four foundations, and it had been a day-and-night grind, unappreciated by

Verrazano-Narrows Bridge Construction
Brooklyn approach · Concreting on upper deck

sidewalk superintendents and, in fact, protested by two hundred Staten Islanders on March 29, 1961; they claimed, in a petition presented to Richmond County District Attorney John M. Braisted, Jr., that the foundation construction between 6 P.M. and 6 A.M. was ruining the sleep of a thousand persons within a one-mile radius. In Brooklyn, the Bay Ridge neighborhood also was cluttered with cranes and earth-moving equipment as work on the approachway to the bridge continued, and the people still were hating Moses, and some had cried foul after he had awarded a $20,000,000 contract, without competitive bidding, to a construction company that employs his son-in-law. All concerned in the transaction immediately denied there was anything irregular about it.

But when Hard Nose Murphy arrived, things were getting better; the bridge was finally crawling up out of the water, and the people had something to see—some visible justification for all the noise at night—and in the afternoons some old Brooklyn men with nothing to do would line the shore watching the robin-red towers climb higher and higher.

The towers had been made in sections in steel plants and had been floated by barge to the bridge site. The Harris Structural Steel Company had made the Brooklyn tower, while Bethlehem made the Staten Island tower—both to O. H. Ammann's specifications. After the tower sections had arrived at the bridge site, they were lifted up by floating derricks anchored alongside the tower piers. After the first three tiers of each tower leg had been locked into place, soaring at this point to about 120 feet, the floating derricks were replaced by "creepers"—derricks, each with a lifting capacity of more than one hundred tons, that crept up the towers on tracks bolted to the sides of the tower legs. As the towers got higher, the creepers were raised until, finally, the towers had reached their pinnacle of 693 feet.

While the construction of towers possesses the element of danger, it is not really much different from building a tall building or an enormous lighthouse; after the third or fourth story is built, it is all the same the rest of the way up. The real art and drama in bridge building begins after the towers are up; then the men have to reach out from these towers and begin to stretch the cables and link the span over the sea.

This would be Murphy's problem, and as he sat in one of

the Harris Company's boats on this morning in May 1962, idly watching from the water as the Staten Island tower loomed up to its tenth tier, he was saying to one of the engineers in the boat, "You know, every time I see a bridge in this stage, I can't help but think of all the problems we got coming next—all the mistakes, all the cursing, all the goddamned sweat and the death we gotta go through to finish this thing . . ."

The engineer nodded, and then they both watched quietly again as the derricks, swelling at the veins, continued to hoist large chunks of steel through the sky.

CHAPTER FIVE

KEEPING THE WHEEL FROM BENNY

After the towers had been finished in the winter of 1962, the cable spinning would begin—and with it the mistakes, the cursing, the sweat, the death that Murphy had anticipated.

The spinning began in March of 1963. Six hundred men were up on the job, but Benny Olson, who had been the best cable-man in America for thirty years, was not among them. He had been grounded. And though he had fumed, fretted, and cursed for three days after he'd gotten the news, it did not help. He was sixty-six years old—too old to be climbing catwalks six hundred feet in the sky, and too slow to be dodging those spinning wheels and snapping wires.

So he was sent four miles up the river to the bridge company's steelyard near Bayonne, New Jersey, where he was made supervisor of a big tool shed and was given some punks to order around. But each day Olson would gaze down the river and see the towers in the distance, and he could sense the sounds, the sights, the familiar sensation that pervades a bridge just before the men begin to string steel thread across the sea. And Benny Olson knew, as did most others, that he had taught the cable experts most of what they knew and had inspired new techniques in the task, and everybody knew, too, that Benny Olson, at sixty-six, was now a legend securely spun into the lore and links of dozens of big bridges between Staten Island and San Francisco.

He was a skinny little man. He weighed about 135 pounds, stood five feet six inches; he was nearly bald on the top of his head, though some strands grew long and loose down the back of his neck, and he had tiny blue eyes, rimmed with steel glasses, and a long nose. Everybody referred to him as "Benny the Mouse." In his long career he had been a pusher, a walkin' boss, and a superintendent. He compensated for his tiny stature by cutting big men down to size, insulting them endlessly and ruthlessly as he demanded perfection and speed on each cable-spinning job. At the slightest provocation he would fire anyone. He would fire his own brother. In fact, he had. On a bridge in Poughkeepsie in 1928, his brother, Ted, did not jump fast enough to one of Benny Olson's commands, and that was all for Ted.

"Now look, you idiots," Olson then told the other men on the bridge, "things around here will be done my way, hear? Or else I'll kick the rest of you the hell off, too, hear?"

Very few men would ever talk back to Benny Olson in those days because, first, they respected him as a bridgeman, as a quick-handed artist who was faster than anybody at pulling wires from a moving wheel and at inspiring a spinning gang to emulate him, and also because Olson, when enraged, was wholly unpredictable and possibly dangerous.

In Philadelphia one day, shortly after he had purchased a new car and was sitting in it at an intersection waiting for the red light to change, a jalopy filled with Negro teenagers came screeching up from behind and banged into the rear bumper of Olson's new car. Quickly, but without saying a word, Olson got out of his car and reached in the back seat for the axe he knew was there. Then he walked back to the boys' car and, still without saying a word, he lifted the axe into the air with both hands and then sent it crashing down upon the fender of the jalopy, chopping off a headlight. Two more fast swings and he had sliced off the other headlight and put a big incision in the middle of the hood. Finally he chopped off a chunk of the aerial with a wide sweep of his axe, and then he turned and walked back to his car and drove slowly away. The boys just sat in their jalopy. They were paralyzed with fear, stunned with disbelief.

Olson was in Philadelphia then because the Walt Whitman Bridge was going up, and the punks hired to work on that bridge were incessantly tormented by Olson, especially the larger ones, and particularly one six-foot two-inch, 235-pound Italian apprentice named Dominick. Every time Benny Olson saw him, he would call him "a dumb bastard" or, at best, a "big, stupid ox."

Just the mere sight of Olson walking down the catwalk would terrorize Dominick, for he was a very high-strung and emotional type, and Olson could get him so nervous and shaky that he could barely light a cigarette. One day, after Olson had hurled five minutes' worth of abuse at Dominick, the big Italian, turning red, lunged toward Olson and grabbed him by the scrawny neck. Then Dominick lifted Olson into the air, carried him toward the edge of the catwalk, and held him out over the river.

"You leetle preek," Dominick screamed, "now I throw you off."

Four other bridgemen rushed up from behind, held Dominick's arms, pulled him back and tried to calm him. Olson, after he'd been let loose, said nothing. He just rubbed his neck and smoothed out his shirt. A moment later he turned and walked idly up the catwalk, but after he had gotten about fifty feet away, Benny Olson suddenly turned and, with a wild flare of fury, yelled to Dominick, "You know, you really are a big, dumb stupid bastard." Then he turned again and continued calmly up the catwalk.

Finally, a few punks on the Walt Whitman Bridge decided to get revenge on Benny Olson. One way to irritate him, they decided, was to stop the spinning wheels, which they could do merely by clicking one of the several turn-off switches installed along the catwalk—placed there in case an accident to one of the men or some flaw in the wiring demanded an instant halt.

So this they did—and, at first, Olson was perplexed. He would be standing on one end of the bridge with everything going smoothly, then, suddenly, a wheel would stop at the other end.

"Hey, what the hell's the matter with that wheel?" he'd yell, but nobody knew. So he would run toward it, running the full length of the catwalk, puffing and panting all the way. Just before he would reach the wheel, however, it would begin to move again—a punk at the other end of the bridge would have flipped the switch back on. This conspiracy went on for hours sometimes, and the game became known as "Keeping the Wheel from Benny." And at 3 A.M. a few punks in a saloon would telephone Benny Olson at his hotel and shout, "Who's got the wheel, Benny?"—and then hang up.

Benny Olson responded without humor, and all day on the bridge he would chase the wheel like a crazy chimpanzee—until, suddenly, he came up with an idea that would stop the game. With help from an engineer, he created an electrical switchboard with red lights on top, each light connected with one of the turn-off switches strung along the bridge. So now if any punk turned off a switch he would give away his location. Olson also appointed a loyal bridge worker to do nothing but watch the switchboard, and this bridgeman was officially called the "tattletale." If the wheel should stop, all Benny Olson had to do was pick up the telephone and say, "Who's got the wheel, Tattletale?" The tattletale would give the precise switch that had been flipped off, and Olson, knowing who was working nearest that spot, could easily fix the blame. But this invention did more than just put an end to the game; it also created a new job in bridge building—the tattletale—and on every big bridge that has been built since the Walt Whitman Bridge, there has been a bridge worker assigned to do nothing but watch the switchboard and keep track of the location of the wheels during the cable-spinning phase of construction. There was a tattletale on the Verrazano-Narrows Bridge, too, but he did little work, for, without Benny Olson to irritate, the demonic spirit had died—there was just no point anymore to "Keeping the Wheel from Benny." And besides, the men involved in spinning the cables on the Verrazano were very serious, very competitive men with no time for games. All they wanted, in the spring of 1963, was to get the catwalks strung up between the towers and the anchorages, and then to get the spinning wheels rolling back and forth across the bridge as quickly and as often as possible. The number of trips that the wheels would make between the anchorages during the daily work-shift of each gang would be recorded in Hard Nose Murphy's of-fice—and it would be a matter of pride for each gang to try to set a daily mark that other gangs could not equal.

Before the spinning could begin, however, the men would have to build a platform on which to stand. This platform would be the two catwalks, each made of wire mesh, each twenty feet wide, each resembling a long thin road of spider web or a mile-long ham-mock. The catwalks would each be held up by twelve horizontal pieces of wire rope, each rope a little more than two inches thick, each more than a mile long. The difficult trick, of course, would be in getting the first of these ropes over the towers of the bridge—a feat that on smaller bridges was accomplished by shooting the rope across with a bow and arrow or, in the case of Charles Ellet's pedes-trian bridge, by paying a boy five dollars to fly a rope across Niagara on the end of a kite.

But with the Verrazano, the first rope would be dragged across the water by barge, then, as the Coast Guard temporarily stopped all ship movements, the two ends of the rope would be hoisted out of the water by the derricks on top of the two towers, more than four thousand feet apart. The other ropes would be hoisted up the same way. Then all would be fastened between the towers, and from the towers back to the anchorages on the extrem-ities of the bridge, following the same "sag" lines that the cables would later follow. When this was done, the catwalk sections would be hauled up. Each catwalk section, as it was lifted, would be folded up like an accordion, but once it had arrived high up on the tower, the bridgemen standing on platforms clamped to the sides of the tower would hook the catwalk sections onto the horizontal ropes, and then shove or kick the catwalk sections forward down the sloping ropes. The catwalks would glide on under the impetus of their own weight and unfurl—as a rolled-up rug might unfurl if pushed down the steep aisle of a movie theatre.

Once all the catwalk sections glided, bumper to bumper, in place, they would be linked end to end, and would be further stiffened by crossbeams. A handrail wire "banister" would also be strung across the catwalks, as would several wooden cross planks to give the men better footing in places where the catwalk was quite steep.

After the two catwalks were in place, another set of wires would be strung above each catwalk, about fifteen feet above, and these upper wires would be the "traveling ropes" that would pull the wheels back and forth, powered by diesel engines mounted atop the anchorages.

Four spinning wheels, each forty-eight inches in diameter and weighing a few hundred pounds apiece, would run simultaneously along the bridge—two wheels atop each of the two catwalks. Each wheel, being double-grooved, would carry two wires at once, and each wheel would take perhaps twelve minutes to cross the entire bridge, averaging eight miles per hour, although it could be speeded up to thirteen miles an hour downhill. As the wheels passed overhead, the men would grab the wires and clamp them down into the specified hooks and pulleys along the catwalk; when a wheel arrived at the anchorage, the men there would remove the wire, hook it in place, reload the wheel and send it back as quickly as possible in the opposite direction.

After the wheel had carried 428 wires across the bridge, the wires would be bound in a strand, and when the wheel had carried across 26,018 wires—or sixty-one strands—they would be squeezed together by hydraulic jacks into a cylindrical shape. This would be a cable. Each cable—there would be four cables on the Verrazano—would be a yard thick, 7,205 feet long, and would contain 36,000 miles of pencil-thin wire. The four cables, collectively, would weigh 38,290 tons. From each cable would later be hung, vertically, 262 suspender ropes—some ropes as long as 447 feet—and they would hold the deck more than two hundred feet above the water, holding it high enough so that no matter how hot and limp the cables got in summer the deck would always be high enough for the Queen Mary to easily pass beneath.

From the very first day that the wheels began to roll—March 7, 1963—there was fierce competition between the two gangs working alongside one another on the two catwalks. This rivalry existed both between the gangs on the early-morning shift as well as the gangs on the late-afternoon shift. The goal of each gang, of course, was to get its two wheels back and forth across the bridge more times than the other gang's wheels. The result was that the cable-spinning operation turned into a kind of horse race or, better yet, a dog race. The catwalks became a noisy arena lined with screaming, fist-waving men, all of them looking up and shouting at their wheels—wheels that became mechanical rabbits.

"Com'on, you mother, move your ass," they yelled as their wheel skimmed overhead, grinding away and carrying the wire to the other end. "Move it, com'on, move it!" And from the other catwalk, there came the same desperate urgings, the same wild-eyed competition and anger when their wheel—their star, their hope—would drag behind the other gang's wheel.

The men from one end of the catwalk to the other were all in rhythm with their wheels, all quick at pulling down the wire, all glancing sideways to study the relative position of the other gang's wheels, all hoping that the diesel engines propelling their wheels would not conk out, all very angry if their men standing on the anchorages were too slow at reloading their wheel once it had completed the journey across. It was in such competition as this that Benny Olson had excelled in his younger days. He used to stand on the catwalk in front of an anchorage inspiring his gang, screaming insults at those too slow at pulling down the wire, or too sluggish at reloading the wheel, or too casual about the competition. Olson was like a deck master hovering over a shipload of slave oarsmen.

On Wednesday, June 19, to the astonishment of the engineers who kept the "score" in Hard Nose Murphy's office, one gang had moved its wheels back and forth across the bridge fifty times. Then, on June 26, a second gang also registered fifty trips. Two days later, in the heat of battle, one of the wheels suddenly broke loose from its moorings and came bouncing down onto the catwalk, skipping toward a bridgeman named John Newberry. He froze with fright. If it hit him, it might knock him off the bridge; if he jumped out of its path too far, he might lose his balance and fall off himself. So he held his position, waiting to see how it jumped. Fortunately, the wheel skimmed by him, he turned slightly like a matador making a pass, and then it stopped dead a few yards down the catwalk. He breathed relief, but his gang was angry because now their daily total was ruined. The other gang would win.

On July 16, one gang got the wheels back and forth fifty-one times, and on July 22, another gang duplicated it. A few days later, the gang under Bob Anderson, the boomer who had been so irresistible to women back on the Mackinac Bridge, was moving along with such flawless precision that with an hour to go of working time it had already registered forty-seven trips. If all went well in the remaining hour, six more trips could be added—meaning a record total of fifty-three.

"Okay, let's move it," Anderson yelled down the line to his gang, all of them focusing on what they hoped would be the winning wheel.

They watched it move smoothly along the tramway overhead, then it rolled higher to the tower, then down, down faster to the anchorage, then up again, quickly reloaded, up the tramway— "Keep moving, you mother!"—closer and closer to the tower now . . . then it stopped.

"Bitch!" screamed one of the punks.

"What the hell's wrong?" shouted Anderson.

"The engine's conked out," some punk finally yelled. "Those goddamn idiots!"

"Let's go beat their asses," yelled another punk, quite serious and ready to run down the catwalk.

"Calm down," Anderson said, with resignation, looking up at the stilled wheel, shaking his head. "Let me go down and see what can be done."

He went down to the anchorage, only to learn that the engine failure could not be fixed in time to continue the race within the hour. So Anderson walked back up, sadly giving the news to his men, and when they walked down the catwalk that night, their hardhats under their arms, their brows sweaty, they looked like a losing football team leaving the field after the game. In the remaining two months, no gang could top the mark of fifty-one, but in September, when the gangs started to place the two-thousand-pound castings over the cables (the castings are metal saddles which would help support the 262 suspender ropes that would stream down vertically from each cable to hold up the deck), a new kind of competition began: a game to see who could bolt into position the most castings, and this got to be dangerous. Not only were bolts dropping off the bridge in this frenzied race—bolts that could pepper the decks of passing vessels and possibly kill anybody they hit—but the castings themselves were unwieldy, and if one of them fell . . .

"Chrissake, Joe, let's get the bolts out and put that mother on," one pusher yelled to Joe Jacklets, who was being cautious with the casting.

The pusher, noticing that another gang working down the catwalk had already removed the bolts and were clamping the casting into place, was getting nervous—his gang was behind.

"Take it easy," Joe Jacklets said, "this thing might not hold."

"It'll hold."

So Joe Jacklets removed the last bolt of the two-section casting and, as soon as he did, one half of the casting—weighing one thousand pounds—toppled off the cable and fell from the bridge.

"Jes-sus !"

"Ohhhhhhh."

"Kee-rist."

"Nooooooooooo !"

"Jes-SUS."

The gang, their hardhats sticking out over the catwalk, watched the one-thousand-pound casting falling like a bomb toward the sea. They noticed, too, a tiny hydrofoil churning through the water below, almost directly below the spot where it seemed the casting might hit. They watched quietly now, mouths open, holding their breaths. Then, after a loud plopping sound, they saw a gigantic splash mushroom up from the water, an enormous fountain soaring forty feet high.

Then, swishing from under the fountain, fully intact, came the hydrofoil, its skipper turning his head away from the splashing spray and shooting his craft in the opposite direction.

"Oh, that lucky little bastard," one of the men said, peering down from the catwalk, shaking his head.

Nobody said anything else for a moment. They just watched the water below. It was as if they hated to turn around and face the catwalk—and later confront Hard Nose Murphy's fiery face and blazing eyes. They watched the water for perhaps two minutes, watched the bubbles subside and the ripples move out. And then, moving majestically into the ripples, moving slowly and peacefully past, was the enormous gray deck of the United States aircraft carrier *Wasp*.

"Holy God!" Joe Jacklets finally said, shaking his head once more.

"You silly bastard," muttered the pusher. Jacklets glared at him.

"What do you mean? I told you it might not hold."

"Like hell you did, you . . ."

Jacklets stared back at the pusher, disbelieving; but then he knew it was no use arguing—he would collect his pay as soon as he could and go back to the union hall and wait for a new job . . .

But before he could escape the scene, the whole line of bridgemen came down the catwalk, some cursing, a few smiling because it was too ridiculous.

"What are you stupid bastards laughing at?" said the walkin' boss.

"Aw, com'on, Leroy," said one of the men, "can't you take a little joke?"

"Yeah, Leroy, don't take it so hard. It's not as if we lost the casting. If we know where a thing is, we ain't lost it."

"Sure, that's right," another said. "We know where it is—it's in the river."

The walkin' boss was just too sick to answer. It was he who would later have to face Murphy.

Across on the other catwalk, the rival gang waved and a few of the younger men smiled, and one yelled out, "Hey, we set ten castings today. How many did you guys set?"

"Nine and a half," somebody else answered.

This got a laugh, but as the workday ended and the men climbed down from the bridge and prepared to invade Johnny's Bar, Joe Jacklets was seen walking with his head down.

If a casting had to fall, it could not have fallen on a better day—September 20, a Friday—because, with work stopped for the weekend anyway, the divers might be able to locate the casting and have it pulled up out of the water before the workers returned to the bridge on Monday. There was no duplicate of the casting, and the plant where it was made was on strike, and so there was no choice but to fish for it—which the divers did, with no success, all day Saturday and Sunday. They saw lots of other bridge parts down there, but no casting. They saw riveting guns, wrenches, and bolts, and there was a big bucket that might have been the one that had fallen with four bolt machines, each worth eight hundred dollars.

Even if it was, the machines as well as the other items were now unserviceable, having been ruined either by the water or the jolt they received when hitting the sea from such high altitudes. Anyway, after a brief inspection of all the tools down there, the divers could easily believe the old saying, "A bridgeman will drop everything off a bridge but money."

Yet this is not precisely true; they drop money off, too. A few five-dollar and ten-dollar bills, even twenty-dollar bills, had been blown off the bridge on some windy Fridays—Friday is pay-day. And during the cable-spinning months, inasmuch as the men were working long hours, they received their pay on the bridge from four clerks who walked along the catwalks carrying more than $200,000 in bundles of cash in zippered camera cases. The cash was sealed in envelopes with each bridgeman's name printed on the outside, and the bridgeman would have to sign a receipt as he received his envelope from the clerk. Some bridgemen, however, after signing the receipt slip, would rip open the envelope and count the money—and that is when they would lose a few bills in the wind. More cautious men would rip off a corner of the envelope, clutching it tight, and count the tips of the bills. Others would just stuff the envelope into their pockets without counting. Still others seemed so preoccupied with their work, so caught up in the competitive swing of spinning, that when the pay clerk arrived with the receipt slip, a pencil, and the envelope, the bridgeman would hastily scribble his name on the slip, then turn away without taking the envelope. Once, as a joke, a clerk named Johnny Cothran walked away with a man's envelope containing more than four hundred dollars, wondering how far he could get with it. He got about twenty feet when he heard the man yelling, "Hey!"

Cothran turned, expecting to face an angry bridgeman. But instead the bridgeman said, "You forgot your pencil." Cothran took the pencil, then handed the bridgeman his envelope. "Thanks," he said, stuffing it absently into his pocket and then quickly getting back to the cable-spinning race.

On Monday, September 23, shortly before noon, the casting was discovered more than one hundred feet below the surface of the narrows, and soon the cranes were swooping over it and pulling it up out of the water. The whole bridge seemed, briefly, to breathe more easily, and Murphy (who had been swearing for three straight days) suddenly calmed down. But two days later, Murphy was again shaking his head in disgust and frustration. At 3:15 P.M. on Wednesday, September 25, somebody on the catwalk had dropped a six-inch steel bolt and, after it had fallen more than one hundred feet, it had hit a bridgeman named Berger Hanson in the face and gone four inches through his skin right under his left eye.

Berger had been standing below the bridge at the time and had been looking up. If he hadn't been looking up the fallen bolt might have hit his hardhat and merely jarred him, instead of doing the damage it did—lifting his eyeball upward, crushing his jawbone, getting stuck in his throat.

Rushed to Victory Memorial Hospital in Brooklyn, Berger was met by the surgeon, Dr. S. Thomas Coppola, who treated all injuries to the bridgemen. Quickly, Dr. Coppola removed the bolt, stopped the bleeding with stitches, then realigned by hand the facial bones and restitched the jaw.

"How do you feel?" Dr. Coppola asked.

"Okay," said Berger.

Dr. Coppola was flabbergasted. "Don't you have any pain?"

"No."

"Can I give you anything—an aspirin or two?"

"No, I'm okay."

After plastic surgery to correct the deformation of his face, and after a few months' recuperation, Berger was back on the bridge.

Dr. Coppola was amazed not only by Berger but by the sto-
icism he encountered in so many other patients among bridgemen.

"These are the most interesting men I've ever met," Dr.
Coppola was telling another doctor shortly afterward. "They're
strong, they can stand all kinds of pain, they're full of pride, and
they live it up. This guy Berger has had five lives already, and he's
only thirty-nine. . . . Oh, I'll tell you, it's a young man's world."

True, the bridge is a young man's world, and old men like Benny
Olson leave it with some bitterness and longing, and hate to be de-
posited in the steelyard on the other side of the river—a yard where
old men keep out of trouble and younger men, like Larry Tatum,
supervise them.

Larry Tatum, a tall, broad-shouldered, daring man of
thirty-seven, had been spotted years ago by Murphy as a "stepper,"
which, in bridge parlance, means a comer, a future leader of
bridgemen.

Tatum had started as a welder when he was only seventeen
years old, and had become a riveter, a fine connector, a pusher. He
had fallen occasionally, but always came back, and had never lost
his nerve or enthusiasm. He had four younger brothers in the business, too—three working under Murphy on the bridge, one having
died under Murphy after falling off the Pan Am building. Larry
Tatum's father, Lemuel Tatum, had been a boomer since the twenties, but now, pushing seventy, he also was in the steelyard, working
under his son, the stepper, watching the boy gain experience as a
walkin' boss so that, quite soon, he would be ready for a promotion
to the number-one job, superintendent.

It was just a little awkward for Larry Tatum, though it was
not obvious, to be ordering around so many old boomers—men

with reputations, like Benny the Mouse, and Lemuel Tatum, and a few dozen others who were in the yard doing maintenance on tools or preparing to load the steel links of the span on barges soon to be floated down to the bridge site. But, excepting for some of Olson's unpredictable explosions, the old men generally were quiet and co-operative—and none more so than the former heavyweight boxing champion, James J. Braddock.

Once they had called Braddock the "Cinderella Man" be-cause, after working as a longshoreman, he won the heavyweight title and earned almost $1,000,000 until his retirement in 1938, after Joe Louis beat him.

Now Braddock was nearly sixty, and was back on the water-front. His main job was to maintain a welding machine. His clothes were greasy, his fingernails black, and his arms so dirty that it was hard to see the tattoos he had gotten one night in the Bowery, in 1921, when he was a frolicsome boy of sixteen.

Now Braddock was earning $170 a week as an oiler, and some men who did not know Braddock might say, as men so often like to say of former champions, "Well, easy come, easy go. Now he's broke, just like Joe Louis."

But his was not another maudlin epic story about a broken prizefighter. Braddock, as he walked slowly around the steelyard, friendly to everyone, his big body erect and his chest out, still was a man of dignity and pride—he was still doing an honest day's work, and this made him feel good.

"What the hell, I'm a working man," he said. "I worked as a longshoreman before I was a fighter, and now I need the money, so I'm working again. I always liked hard work. There's nothing wrong with it."

He lost $15,000 on a restaurant, Braddock's Corner, once on

71

West Forty-ninth Street in Manhattan, and the money he had put into a marine supply house, which he operated for ten years, proved not to be a profitable venture. But he still owns the $14,000 home he bought in North Bergen, New Jersey, shortly after the Joe Louis fight, he said, and he still loves his wife of thirty-three years' marriage, and still has his health and a desire to work hard, and has two sons who work hard, too.

One son, Jay, who is thirty-two, weighs 330 pounds and stands six feet five inches. He works in a Jersey City powerhouse; the other, thirty-one-year-old Howard, is a 240-pounder who is six feet seven inches and is in road construction. "So don't feel sorry for me," James J. Braddock, the former Cinderella Man, said, inhaling on a cigarette and leaning forward on a big machine. "Don't feel sorry for me one bit."

But he did admit that bridge building, like boxing, was a young man's game.

And of all the eager young men working on the Verrazano-Narrows Bridge under Hard Nose Murphy in the fall of 1963, few seemed better suited to the work or happier on a bridge job than the two men working together atop the cable 385 feet over the water behind the Brooklyn tower.

One was very small, the other very large. The small man, standing five feet seven inches and weighing only 138 pounds—but very sinewy and tough—was named Edward Iannielli. He was called "The Rabbit" by the other men because he jumped the beams and ran across wires, and everybody said of the twenty-seven-year-old Iannielli that he would never live to be thirty.

The big boy was named Gerard McKee. He was a handsome, wholesome boy, about two hundred pounds and six feet three and one-half inches. He had been a Coney Island lifeguard, had charm with women and a gentle disposition, and all the men on the bridge immediately took to him, although he was not as friendly and forward as Iannielli.

On Wednesday morning, October 9, the two climbed the cables as usual, and soon, amid the rattling of the riveters and clang of mallets, they were hard at work, heads down, tightening cable bolts, barely visible from the ground below.

Before the morning was over, however, the attention of the whole bridge would focus on them.

DEATH ON A BRIDGE

It was a gray and windy morning. At 6:45 A.M. Gerard McKee and Edward Iannielli left their homes in two different parts of Brooklyn and headed for the bridge.

Iannielli, driving his car from his home in Flatbush, got there first. He was already on the catwalk, propped up on a cable with one leg dangling 385 feet above the water, when Gerard McKee walked over to him and waved greetings.

The two young men had much in common. Both were the sons of bridgemen, both were Roman Catholics, both were natives of New York City, and both were out to prove something—that they were as good as any boomer on the bridge.

They quietly resented the prevailing theory that boomers

make the best bridgemen. After all, they reasoned, boomers were created more out of necessity than desire; the Indian from the reservation, the Southerner from the farm, the Newfoundlander from the sea, the Midwesterner from the sticks—those who composed most of the boomer population—actually were escaping the poverty and boredom of their birthplaces when they went chasing from boom town to boom town. Iannielli and McKee, on the other hand, did not have to chase all over America for the big job; they could wait for the job to come to them, and did, because the New York area had been enjoying an almost constant building boom for the last ten years.

And yet both were impressed with the sure swagger of the boomer, impressed with the fact that boomers were hired on jobs from New York to California, from Michigan to Louisiana, purely on their national reputations, not on the strength of strong local unions.

This realization seemed to impress Iannielli a bit more than McKee. Perhaps it was partly due to Iannielli's being so small in this big man's business.

He, like Benny Olson, desperately wanted to prove himself, but he would make his mark not by cutting big men to size, or by boasting or boozing, but rather by displaying cold nerve on high steel—taking chances that only a suicidal circus performer would take—and by also displaying excessive pride on the ground.

Iannielli loved to say, "I'm an ironworker." (Bridges are now made of steel, but iron was the first metal of big bridges, and the first bridgemen were called "ironworkers." There is great tradition in the title, and so Iannielli—and all bridgemen with pride in the past—refer to themselves as ironworkers, never steelworkers.)

When Edward Iannielli first became an apprentice iron-worker, he used to rub orange dust, the residue of lead paint, into his boots before taking the subway home; he was naive enough in those days to think that passengers on the subway would associate orange dust with the solution that is coated over steel during construction to make it rustproof.

"When I was a little kid growing up," he had once recalled, "my old man, Edward Iannielli, Sr., would bring other ironworkers home after work, and all they'd talk about was ironwork, ironwork. That's all we ever heard as kids, my brother and me. Sometimes my old man would take us out to the job, and all the other ironworkers were nice to us because we were Eddie's sons, and the foreman might come over and ask, 'You Eddie's sons?' and we'd say, 'Yeah,' and he'd say, 'Here, take a quarter.' And that is how I first started to love this business.

"Later, when I was about thirteen or fourteen, I remember going out to a job with the old man and seeing this big ladder. And I yelled to my father, 'Can I climb up?' and he said, 'Okay, but don't fall.' So I began to climb up this thing, higher and higher, a little scared at first, and then finally I'm on the top, standing on this steel beam way up there, and I'm all alone and looking all around up there, looking out and seeing very far, and it was exciting, and as I stood up there, all of a sudden, I am thinking to myself, 'This is what I want to do!'"

After his father had introduced him to the business agent of Local 361, the ironworkers' union in Brooklyn, Edward Iannielli, Jr., started work as an apprentice.

"I'll never forget the first day I walked into that union hall," he had recalled. "I had on a brand-new pair of shoes, and I saw all those big men lined up, and some of them looked like bums, some looked like gangsters, some just sat around tables playing cards and cursing.

"I was scared, and so I found a little corner and just sat there, and in my pocket I had these rosaries that I held. Then a guy walked out and yelled, 'Is young Iannielli here?' and I said, 'Here,' and he said, 'Got a job for you.' He told me to go down and report to a guy named Harry at this new twelve-story criminal court building in downtown Brooklyn, and so I rushed down there and said to Harry, 'I'm sent out from the hall,' and he said, 'Oh, so you're the new apprentice boy,' and I said, 'Yeah,' and he said, 'You got your parents' permission?' and I said, 'Yeah,' and he said, 'In writing?' and I said, 'No,' and so he said, 'Go home and get it.'

"So I get back on the subway and go all the way back, and I remember running down the street, very excited because I had a job, to get my mother to sign this piece of paper. Then I ran all the way back, after getting out of the subway, up to Harry and gave him the piece of paper, and then he said, 'Okay, now I gotta see your birth certificate.' So I had to run all the way back, get another subway, and then come back, and now my feet in my new shoes are hurting.

"Anyway, when I gave Harry the birth certificate, he said, 'Okay, go up that ladder and see the pusher,' and when I got to the top, a big guy asked, 'Who you?' I tell 'im I'm the new apprentice boy, and he says, 'Okay, get them two buckets over there and fill 'em up with water and give 'em to the riveting gang.

"These buckets were two big metal milk cans, and I had to carry them down the ladder, one at a time, and bring them up, and this is what I did for a long time—kept the riveting gangs supplied with drinking water, with coffee and with rivets—no ifs and buts, either.

"And one time, when I was on a skyscraper in Manhattan, I remember I had to climb down a ladder six floors to get twenty coffees, a dozen sodas, some cake and everything, and on my way back, holding everything in a cardboard box, I remember slipping on a beam and losing my balance. I fell two flights. But luckily I fell in a pile of canvas, and the only thing that happened was I got splashed in all that steaming hot coffee. Some ironworker saw me laying there and he yelled, 'What happened?' and I said, 'I fell off and dropped the coffee,' and he said, '*You dropped the coffee!* Well, you better get the hell down there fast, boy, and get some more coffee.'

"So I go running down again, and out of my own money— must have cost me four dollars or more—I bought all the coffee and soda and cake, and then I climbed back up the ladder, and when I saw the pusher, before he could complain about anything, I told him I'm sorry I'm late."

After Edward Iannielli had become a full-fledged iron-worker, he fell a few more times, mostly because he would run, not walk across girders, and once—while working on the First National City Bank in Manhattan—he fell backward about three stories and it looked as if he was going down all the way. But he was quick, light and lucky—he was "The Rabbit," and he landed on a beam and held on.

"I don't know what it is about me," he once tried to explain, "but I think it all has something to do with being young, and not wanting to be like those older men up there, the ones that keep telling me, 'Don't be reckless, you'll get killed, be careful.' Sometimes, on windy days, those old-timers get across a girder by crawling on their hands and knees, but I always liked to run across and show those other men how to do it. That's when they all used to say, 'Kid, you'll never see thirty.'

"Windy days, of course, are the hardest. Like you're walk-ing across an eight-inch beam, balancing yourself in the wind, and then, all of a sudden, the wind stops—and you temporarily lose your balance. You quickly straighten out—but it's some feeling when that happens."

Edward Iannielli first came to the Verrazano-Narrows job in 1961, and while working on the Gowanus Expressway that cut through Bay Ridge, Brooklyn, to the bridge, he got his left hand caught in a crane one day.

One finger was completely crushed, but the other, cleanly severed, remained in his glove. Dr. Coppola was able to sew it back on. The finger would always be stiff and never as strong as before, of course; yet the surgeon was able to offer Edward Iannielli two choices as to how the finger might be rejoined to his hand. It could either be set straight, which would make it less conspicuous and more attractive, or it could be shaped into a grip-form, a hook. While this was a bit ugly, it would mean that the finger could more easily be used by Iannielli when working with steel. There was no choice, as far as Iannielli was concerned; the finger was bent permanently into a grip.

When, in the fall of 1963, Gerard McKee met Edward Iannielli and saw the misshapen left hand, he did not ask any questions or pay any attention. Gerard McKee was a member of an old family of construction workers, and to him malformation was not uncommon, it was almost a way of life. His father, James McKee, a big, broad-shouldered man with dark hair and soft blue eyes—a man whom Gerard strongly resembled—had been hit by a collapsing crane a few years before, had had his leg permanently twisted, had a steel plate inserted in his head, and was disabled for life.

James McKee had been introduced to ironwork by an uncle, the late Jimmy Sullivan, who had once been Hard Nose Murphy's boss in a gang. The McKee name was well known down at Local 40, the union hall in Manhattan, and it had been quite logical for James McKee, prior to his accident, to take his three big sons down to the hall and register them in the ironworkers' apprentice program.

Of the three boys, Gerard McKee was the youngest, tallest and heaviest—but not by much. His brother John, a year older than Gerard, was 195 pounds and six feet two inches. And his brother Jimmy, two years older than Gerard, was 198 pounds and six feet three inches.

When the boys were introduced to union officials of Local 40, there were smiles of approval all around, and there was no doubt that the young McKees, all of them erect and broad-shouldered and seemingly eager, would someday develop into superb ironworkers. They looked like fine college football prospects—the type that a scout would eagerly offer scholarships to without asking too many embarrassing questions about grades. Actually, the McKee boys had never even played high school football. Somehow in their neighborhood along the waterfront of South Brooklyn, an old Irish neighborhood called Red Hook, the sport of football had never been very popular among young boys.

The big sport in Red Hook was swimming, and the way a young boy could win respect, could best prove his valor, was to jump off one of the big piers or warehouses along the waterfront, splash into Buttermilk Channel, and then swim more than a mile against the tide over to the Statue of Liberty.

Usually, upon arrival, the boys would be arrested by the guards. If they weren't caught, they would then swim all the way back across Buttermilk Channel to the Red Hook side.

None of the neighborhood boys was a better swimmer than Gerard McKee, and none had gotten back and forth through

Buttermilk Channel with more ease and speed than he. All the young boys of the street respected him, all the young girls who sat on the stoops of the small frame houses admired him—but none more than a pretty little Italian redhead named Margaret Nucito, who lived across the street from the McKees.

She had first seen Gerard in the second grade of the parochial school. He had been the class clown—the one the nuns scolded the most, liked the most.

At fourteen years of age, when the neighborhood boys and girls began to think less about swimming and more about one another, Margaret and Gerard started to date regularly. And when they were eighteen they began to think about marriage. In the Red Hook section of Brooklyn, the Catholic girls thought early about marriage. First they thought about boys, then the Prom, then marriage. Though they thought Hook was a poor neighborhood of shanties and small two-story frame houses, it was one where engagement rings were nearly always large and usually expensive. It was marriage before sex in this neighborhood, as the Church preaches, and plenty of children; and, like most Irish Catholic neighborhoods, the mothers usually had more to say than the fathers. The mother was the major moral strength in the Irish church, where the Blessed Virgin was an omnipresent figure; it was the mother who, after marriage, stayed home and reared the children, and controlled the family purse strings, and chided the husband for drinking, and pushed the sons when they were lazy, and protected the purity of her daughters.

And so it was not unusual for Margaret, after they tentatively planned marriage and after Gerard had begun work on the bridge, to be in charge of the savings account formed by weekly deductions from his ironworker's earnings. He would only fritter the money away if he were in control of it, she had told him, and he did not disagree. By the summer of 1965 their account had reached $800. He wanted to put this money toward the purchase of the beautiful pear-shaped diamond engagement ring they had seen one day while walking past Kastle's jewelry window on Fulton Street. It was a one-and-one-half carat ring priced at $1,000. Margaret had insisted that the ring was too expensive, but Gerard had said, since she had liked it so much, that she would have it. They planned to announce their engagement in December.

On Wednesday morning, October 9, Gerard McKee hated to get out of bed. It was a gloomy day and he was tired, and downstairs his brothers were yelling up to him, "Hey, if you don't get down here in two minutes, we're leaving without you." He stumbled down the steps. Everyone had finished breakfast and his mother had already packed three ham-and-cheese sandwiches for his lunch. His father, limping around the room, was quietly cross at his tardiness.

He had not been out late the night before. He had gone over to Margaret's briefly, then had had a few beers at Gabe's, a neighborhood saloon with a big bridge painted across the back-bar. He had been in bed by about midnight, but this morning he ached, and he suspected he might be getting a cold.

They all left the house at 6:45 A.M. and caught a bus near the corner; then at Forty-ninth Street they got a cab and rode it to One Hundred and First Street in Bay Ridge, and then they walked, with hundreds of other ironworkers, down the dirt road toward the Brooklyn anchorage of the Verrazano–Narrows Bridge.

"Wait, let me grab a container of coffee," Gerard said, stopping at a refreshment shack along the path.

"Hurry up."

"Okay," he said. He gulped down the coffee in three swallows as they walked, and then they all lined up to take the elevator up to the catwalk. That morning Jimmy and John McKee were working on the section of the catwalk opposite Gerard, and so they parted on top, and Gerard said, "See you tonight." Then he headed off to join Iannielli.

Edward Iannielli seemed his spry self. He was sitting up there on the cable, whistling and very chipper.

"Good morning," he said, and McKee waved and forced a smile. Then he climbed up on the cable, and soon they began to tighten the seven bolts on the top of the casting.

After they had finished, Iannielli slid down from the cable and McKee handed the ratchet wrench down to him. Iannielli then fitted the wrench to the first of the seven lower bolts.

It was now about 9:30 A.M. It was cloudy and windy, though not as windy as it had been in the first week of October. Iannielli pushed his hardhat down on his head. He gazed down the catwalk and could see hundreds of men, their khaki shirts and jackets billowing in the breeze, all working on the cable—bolting it, banging it with tools, pushing into it. Iannielli took the big wrench he held, fixed it to a bolt, and pressed hard. And then, suddenly, from the bottom of the cable he heard a voice yelling, "Eddie, Eddie . . . help me, Eddie, help me . . . please, Eddie . . ."

Iannielli saw, hanging by his fingers from the south edge of the catwalk, clutching tightly to the thin lower wires of the hand rail, the struggling figure of Gerard McKee.

"God," Iannielli screamed, "dear God," he repeated, lunging forward, lying across the catwalk and trying to grab onto McKee's arms and pull him up. But it was very difficult.

Iannielli was only 138 pounds, and McKee was more than

200. And Iannielli, with one finger missing on his left hand from the crane accident, and with the resewn second finger not very strong, could not seem to pull McKee's heavy body upward even one inch. Then McKee's jacket and shirt came loose, and he seemed to be just hanging there, dead weight, and Iannielli kept pleading, "Oh, God, God, please bring him up . . . bring him up . . ."

Other men, hearing the screams, came running and they all stretched down, grabbing wildly for some part of McKee's clothing, and Gerard kept saying, "Hurry, hurry please, I can't hold on any longer." And then, a few moments later, he said, "I'm going to go . . . I'm going to go . . ." and he let go of the wire and dropped from the bridge.

The men watched him fall, feet first for about one hundred feet. Then his body tilted forward, and Iannielli could see McKee's shirt blowing off and could see McKee's bare back, white against the dark sea, and then he saw him splash hard, more than 350 feet below, and Iannielli closed his eyes and began to weep, and then he began to slip over, too, but an Indian, Lloyd LeClaire, jumped on top of him, held him tight to the catwalk.

Not far from where Gerard McKee hit the water, two doctors sat fishing in a boat, and also nearby was a safety launch. And for the next thirty seconds, hysterical and howling men's voices, dozens of them, came echoing down from the bridge, "Hey, grab that kid, grab that kid . . . hurry, grab that kid . . ."

Even if Gerard McKee had landed within a yard of the safety boat, it would have been no use; anyone falling from that altitude is sure to die, for, even if his lungs hold out, the water is like concrete, and bodies break into many pieces when they fall that far.

The remains of Gerard McKee were taken out of the water and put into the safety launch and taken to Victory Memorial

Hospital. Some of the men up on the bridge began to cry, and, slowly, all of them, more than six hundred of them, removed their hardhats and began to come down. Work was immediately suspended for the day. One young apprentice ironworker, who had never seen a death like this before, froze to the catwalk and refused to leave; he later had to be carried down by three others.

Jimmy and John McKee went home to break the news and be with their parents and Margaret, but Edward Iannielli, in a kind of daze, got into his automobile and began to drive away from the bridge, without any destination. When he saw a saloon he stopped. He sat at the bar between a few men, shaking, his lips quivering. He ordered one whiskey, then another, then three beers. In a few minutes he felt loose, and he left the bar and got into his car and began to drive up the Belt Parkway. He drove about fifty miles, then, turning around, he drove fifty miles back, seeing the bridge in the distance, now empty and quiet. He turned off the Belt Parkway and drove toward his home. His wife greeted him, very excitedly, at the door, saying that the bridge company had called, the safety officer had called, and what had happened?

Iannielli heard very little of what she was saying. That night in bed all he could hear, over and over, was "Eddie, Eddie, help me . . . help me." And again and again he saw the figure falling toward the sea, the shirt blowing up and the white back exposed. He got out of bed and walked through the house for the rest of the night.

The next day, Thursday, October 10, the investigation was begun to determine the cause of McKee's death. Work was again suspended on the bridge. But since nobody had seen how McKee had gotten off the cable, nobody knew whether he had jumped onto the catwalk and bounced off it or whether he had tripped—and they still do not know. All they knew then was that the morale of the men was shot, and Ray Corbett, business agent for Local 40, began a campaign to get the bridge company to string nets under the men on the bridge.

This had not been the first death around the bridge. On August 24, 1962, one man fell off a ladder inside a tower and died, and on July 13, 1963, another man slipped off the approach road and died. But the death of Gerard McKee was somehow different— different, perhaps, because the men had watched it, had been helpless to stop it; different, perhaps, because it had involved a very popular young man, the son of an ironworker who had himself been crippled for life.

Whatever the reason, the day of Gerard McKee's death was the blackest day on the bridge so far. And it would have made little difference for any company official to point out that the Verrazano-Narrows' safety record—just three deaths during thousands of working hours involving hundreds of men—was highly commendable.

McKee's funeral, held at the Visitation Roman Catholic Church in Red Hook, was possibly the largest funeral ever held in the neighborhood. All the ironworkers seemed to be there, and so were the engineers and union officials. But of all the mourners, the individual who seemed to take it the worst was Gerard's father, James McKee.

"After what I've been through," he said, shaking his head, tears in his eyes, "I should know enough to keep my kids off the bridge."

CHAPTER SEVEN

STAGE IN THE SKY

Gerard McKee's two brothers quit the bridge immediately, as their father had requested, but both were back within the month. The other ironworkers were a bit nervous when the McKees climbed up that first day back, but the brothers assured everyone that it was far more comfortable working up on the bridge, busy among the men, than remaining in the quietude of a mournful home.

      Though nobody could ever have imagined it then, the death of Gerard McKee was just the beginning of a long, harsh winter— possibly the worst in Hard Nose Murphy's career. There would be a tugboat strike and a five-day ironworkers' strike to force management to put nets under the bridge; there would be freezing weather, powerful winds that would swing the bridge, careless mistakes that would result in a near disaster while the men were lifting a four-hundred-ton piece of steel; and, hovering over everything else, there would be the assassination on November 22 of President Kennedy, an event that demoralized men nowhere in the world more than it did on the bridge, where the majority of workers were of Irish ancestry.

All of this would occur while Hard Nose Murphy and the American Bridge engineers were facing their greatest challenge—the span across the sea.

If construction was to remain on schedule, permitting the bridge to open in late November of 1964, then the steel skeleton of the span would have to be linked 6,690 feet across the sky by spring—a feat that now, in the winter of 1963, seemed quite impossible.

The task would involve the hoisting off barges of sixty separate chunks of steel, each the size of a ten-room ranch house (but each weighing four hundred tons), more than 220 feet in the air. Each of these steel pieces, in addition to several smaller ones, would then be linked to the suspender ropes dangling from the cables and would finally be locked together horizontally across the water between Brooklyn and Staten Island.

If one of these pieces dropped, it would set the bridge's schedule back at least six months, for each piece was without a duplicate. The sixty larger pieces, all of them rectangular in shape, would be about twenty-eight feet high, 115 feet wide, almost as long, and would be floated to the bridge, one at a time, from the American Bridge Company's steelyard four miles up the river in New Jersey, where Benny Olson, James Braddock, and the other old champs were working. The loaded barges, pulled by tugboats, would take an hour to make the trip. Once the steel pieces were lifted off the barges by two tremendous hoisting machines on the lower traverse strut of each tower, the whole bridge would sag under the pressure of weight; for instance, the first piece, when lifted up, would pull the main cables down twenty inches. The second and third pieces would lower the cables an additional four feet six inches. The fifth and sixth pieces would pull the cables down another four feet three inches. When all the pieces were hanging, the cables would be as much as twenty-eight feet lower than before. (All this was as O. H. Ammann had designed it—in fact, his design allowed for as much as a thirty-five-foot cable defection—but he did not take into account human and mechanical frailty, this being Murphy's problem.)

Murphy's problems did not begin with the lifting of the first few steel pieces. This was conducted in the presence of a boatload of television and news cameras, and all the workers were very much on the ball. His troubles began when the initial excitement was tempered by the rote of repetition and the coming of colder weather. One freezing day a small barge holding suspender ropes was tied too tightly to the pier and sank that night when the tide came in, and the guard not only slept through this but also permitted vagrants to ransack the tool shed.

Murphy, in his shack the next morning, pounding his fist against the desk, was on the phone screaming to one of the dock supervisors. "Jes-sus Kee-rist, I'm sick of this crap! That stupidbastardguard just stood in that warm shanty, sleeping instead of watching. Now that guard isn't supposed to be sleeping where it's warm, goddammit, he's supposed to be watching, and I'm not taking any more of this crap, so you get that goddam guard up here and I'll tell that stupid bastard a thing or two . . ."

In the outer office of the shack, Murphy's male secretary, a slim, dapper, well-groomed young man named Chris Reisman, was on the switchboard answering calls with a very polite, "Good morning, American Bridge" and covering his ears to Murphy's profanity in the next room.

Male secretaries are the only sort that would survive in this atmosphere; a female secretary would probably not be safe around some of the insatiable studs who work on bridges, nor would any woman condone the language very long. But Chris Reisman, whose uncle was a riveter and whose stepfather died on a bridge six years before, worked out well as a secretary, although it took a while for the bridgemen to accustom themselves to Chris Reisman's polite telephone voice saying, "Good morning, American Bridge" (instead of "Yeah, whatyawant?"), and to his style of wearing slim, cuffless trousers, a British kneelength raincoat, and, sometimes in wet weather, high soft leather boots.

The day after Reisman had been hired by the American Bridge Company and sent to Murphy's shack on the Staten Island shore, Murphy's welcoming words were, "Well, I see we got another ass to sit around here." But soon even Murphy was impressed with twenty-three-year-old Reisman's efficiency as a secretary and his cool manner over the telephone in dealing with people Murphy was trying to avoid.

"Good morning, American Bridge . . ."

"Yeah, say, is Murphy in?"

"May I ask who's calling?"

"Wha?"

"May I ask who's calling?"

"Yeah, dis is an old friend, Willy . . . just tell 'im Willy . . ."

"May I have your last name?"

"Wha?"

"Your last name?"

"Just tell Murphy, well, maybe you can help me. Ya see, I worked on the Pan Am job with Murphy, and . . ."

"Just a minute, please," Chris cut in, then switched to Murphy on the intercom and said, "I have a Willy on the phone that worked for you . . ."

"I don't want to talk to that bastard," Murphy snapped back.

Then, back on the phone, Reisman said, "I am sorry, sir, but Mr. Murphy is not in."

"Wha?"

"I said I do not expect Mr. Murphy to be in today."

"Well, okay, I'll try tomorrow."

"Fine," said Chris Reisman, clicking him off, then picking up another call with, "Good morning, American Bridge . . ."

On Thursday, November 21, there was a hoisting engine failure, and a four-hundred-ton steel unit, which was halfway up, could not go any farther, so it dangled there all night. The next day, after the engine trouble was corrected, there was union rumbling over the failure of the bridge company to put nets under the bridge. This fight was led by Local 40's business agent, Ray Corbett, himself a onetime ironworker—he helped put up the television tower atop the Empire State—and on Monday, December 2, the men walked off the bridge because of the dispute.

The argument against nets was not so much money or the time it would take to string them up, although both of these were factors, but mainly the belief that nets were not really a safeguard against death. Nets could never be large enough to cover the whole underbelly of the bridge, the argument went, because the steel had to be lifted up into the bridge through the path of any nets. It was also felt that nets, even small ones strung here and there, and moved as the men moved, might induce a sense of false security and invite more injury than might otherwise occur.

The strike lasted from December 2 to December 6, ending with the ironworkers' unions victorious—they got their nets, small as they seemed, and Ray Corbett's strong stand was largely vindicated within the next year, when three men fell off the bridge and were saved from the water by dropping into nets. By January, with barges arriving every day with one and sometimes two four-hundred-ton steel pieces to be lifted, about half of the sixty box-shaped units were hanging from the cables, and things seemed, at least for the time being, to be under control. Each day, if the sun was out, the old bridge buffs with binoculars would shiver on the Brooklyn shoreline, watching and exchanging sage comments and occasionally chatting with the ironworkers who passed back and forth through the gate with its sign

BEER OR ANY ALCOHOLIC BEVERAGES NOT PERMITTED ON THIS JOB. APPRENTICES WHO BRING ALCOHOLIC BEVERAGES ON JOB FOR THE MEN WILL BE TERMINATED.

"You never drink on the bridge, right?" a man near the gate asked an Indian ironworker named Bronco Bill Martin.

"Who?"

"You."

"No, I only drink beer."

"Well, doesn't beer ruin your sense of balance?"

"I donno," Bronco Bill said. "I just go to job, drink beer, climb bridge, and I feel better on bridge than I do on ground. I can drink a dozen can of beer and still walk a straight line on that bridge."

"A dozen?"

"Yeah," he said, "easy."

A few yards away, a group of white-haired men, some of

Verrazano-Narrows Bridge Construction   Brooklyn side span during spinning operation

them retired engineers or construction workers, all of them now "seaside superintendents," were peering up at the bridge, listening to the grinds of the hoisting machines and the echoes of "Red" Kelly shouting instructions up through his bullhorn from a barge below the rising four-hundred-ton steel unit. It was a fascinating water show, very visual and dramatic, even for these elderly men who only saw the finale.

The show had its beginning more than an hour earlier up the river on the Jersey side. There, along the waterfront of the American Bridge Company's yard, a four-hundred-ton chunk of steel (steel that had been made in smaller sections at U.S. Steel plants in other states, and then shipped by rail to the New Jersey yard for assembly) was resting on a gigantic twin barge and was now just being pulled away by one tug, pushed by another.

The ironworkers, about seventy of them, waved from the yard to the tugs—another four hundred tons was off. The tug pilot, a thin, blond Norwegian-American named Villy Knutsen, carefully churned through New York Harbor—crisscrossing with oil tankers, ferryboats, luxury cruisers, aircraft carriers, fishing boats, driftwood, floating beer cans—squinting his pale blue eyes in the sun and splashing spray. On this day he was talking to a deckhand, Robert Guerra, telling him how he had hated the bridge when it first began. The Knutsens had been one of the families in Bay Ridge whose homes had been threatened by the approaches to the bridge. Villy Knutsen and his wife had joined the protests and signed many petitions before finally moving to Port Jefferson, Long Island.

"But I really hated that bridge then, I'll tell ya," he repeated.

"Well, don't hate it anymore," Guerra said, "it's making you a day's pay."

"Yeah," agreed Knutsen, cutting his tugboat wheel around quickly to skirt past an oncoming tanker, then swinging his head around to observe the barges still slapping the sea under the big red steel; all was well.

Forty minutes later, Knutsen was bringing his tug with its

big steel caboose toward the bridge. The old men on the shore lifted their binoculars, and the ironworkers on the bridge got ready, and under one of the towers a fat pusher, with a telephone pressed to his left ear, was looking up at the crane on the tower and frowning and saying, "Hello, Eddie? Eddie? Hello, Eddie?" Eddie, the signal man on top, did not answer.

"Hello, Eddie?"

Still no answer.

"Gimme that phone," said another pusher, grabbing it. "Hello, Eddie? Hello, Eddie?"

"Hello," came a thin voice through the static.

"Hello, Eddie?"

"No, Burt."

"Burt, this is Joe down the bottom. Eddie in the cab?"

Silence.

"Hello, Burt? Hello, Burt?" . . . "Christsakes!" the pusher with the phone shouted, holding it away from his ear and frowning at it again. Then he put it to his mouth again. "Hello, Burt? Burt, hello, Burt . . . ?"

"Yeah."

"Whatthehell's wrong, Burt?"

"You got a goddam broken splice in your hand down there."

"Well, keep talking, Burt."

"Okay, Joe."

"Keep talking, Burt . . . keep . . ." The phone went dead again.

"Christsakes," Joe shouted. "Hello, Burt? Hello? Hellohellohello? Nuthin' . . . Hello? . . . Nuthin'. Hello, Burt?"

"Hello," came the voice from the top.

"Burt?"

"No, this is Eddie."

"Keep talkin', Eddie . . ."

Finally the telephone system between the pusher on the ground and the man on top who controlled the movements of the crane was re-established. Soon Villy Knutsen's tug had bumped the barge into position, and the next step was to link the hauling ropes to the steel piece and prepare to hoist it from the barge up 225 feet to the span.

This whole operation would be under the direct supervision of one man, who now stepped out of a workboat onto the barge— onto the stage. He was a big, barrel-chested man of 235 pounds who stood six feet two inches, and he was very conspicuous, even from the shore, because he wore a red checkered jacket and a big brown hardhat tilted forward over his red hair and big ruddy nose, and because he carried in his right hand a yellow bullhorn through which his commands could be heard by all the men on top of the bridge. He was Jack "Red" Kelly, the number-two bridgeman, second only to Hard Nose Murphy himself.

Suddenly all the men's attention, and also the binoculars along the shore, were focused on Kelly as he carefully watched a dozen ironworkers link the heavy hoisting ropes to the four upper extremities of the steel piece as it gently rocked with the anchored barge. When the ropes had been securely bound, the men jumped off the barge onto another barge, and so did Kelly, and then he called out, "Okay, ease now . . . up . . . UP . . ."

Slowly the hauling machines on the tower, their steel thread strung up through the cables and then down all the way to the four edges of the steel unit on the sea, began to grind and grip and finally lift the four hundred tons off the barge.

Within a few minutes, the piece was twenty feet above the

barge, and Kelly was yelling, "Slack down . . ." Then, "Go ahead on seven," "Level it up," "Go ahead on seven"; and up on the bridge the signal man, phones clamped to his ears, was relaying the instructions to the men inside the hauling vehicles. Within twenty-five minutes, the unit had climbed 225 feet in the air, and the connectors on top were reaching out for it, grabbing it with their gloves, then linking it temporarily to one of the units already locked in place.

"Artfully done," said one of the old men with binoculars. "Yes, good show," said another.

Most of the sixty pieces went up like this one—with remarkable efficiency and speed, despite the wind and bitter cold, but just after New Year's Day, a unit scheduled to be lifted onto the Brooklyn backspan, close to the shoreline, caused trouble, and the "seaside superintendents" had a good view of Murphy's temper.

As the steel piece was lifted a few feet off the barge, a set of hold-back lines that stretched horizontally from the Brooklyn tower to the rising steel unit broke loose with a screech. (The hold-back lines were necessary in this case because the steel unit had to rise at an oblique angle, the barge being unable to anchor close enough to the Brooklyn shoreline to permit the unit a straight ride upward.)

Suddenly, the four-hundred-ton steel frame twisted, then went hurtling toward the Brooklyn shore, still hanging on ropes, but out of control. It swished within a few inches of a guard-rail fence beneath the anchorage, then careened back and dangled uneasily above the heads of a few dozen workmen. Some fell to the ground; others ran.

"Jes-sus," cried one of the spectators on shore. "Did you see that?"

"Oh, My!"

"Oh, if old man Ammann were here now, he'd have a fit!"

Down below, from the pier and along the pedestals of the tower, as well as up on the span itself, there were hysterical cries and cursing and fist waving. A hurried telephone call to Hard Nose Murphy brought him blazing across the Narrows in a boat, and his swearing echoed for a half-hour along the waterfront of Bay Ridge.

"Which gang put the clamps on that thing?" Murphy demanded.

"Drilling's gang," somebody said.

"And where the hell's Drilling?"

"Ain't here today."

Murphy was probably never angrier than he was on this particular Friday, January 3. John Drilling, who had led such a charmed life back on the Mackinac Bridge in Michigan, and who had recently been promoted to pusher though he was barely twenty-seven years old, had called in sick that day. He was at his apartment in Brooklyn with his blond wife, a lovely girl whom he'd met while she was working as a waitress at a Brooklyn restaurant. His father, Drag-Up Drilling, had died of a heart attack before the cable spinning had begun, and some people speculated that if he had lived he might have been the American Bridge superintendent in place of Hard Nose Murphy. The loss of his father and the responsibility of a new wife and child had seemed to change John Drilling from the hell-raiser he had been in Michigan to a mature young man. But now, suddenly, he was in real trouble. Though he was not present at the time of the accident that could have caused a number of deaths, he was nevertheless responsible; he should have checked the clamps when they were put on by his gang the day before.

When John Drilling returned to the bridge, not yet aware of the incident, he was met by his friend and fellow boomer, Ace Cowan, a big Kentuckian who was the walkin' boss along the Brooklyn backspan.

"What you got lined out for today, Ace?" Drilling said cheerfully upon arrival in the morning.

"Er, well," Cowan said, looking at his feet, "they . . . the office made me put you back in the gang."

"The gang! What, for taking a lousy day off?"

"No, it was those clamps that slipped . . ."

Drilling's face fell.

"Anybody hurt?"

"No," Cowan said, "but everybody is just pissed . . . I mean Ammann and Whitney, and Murphy, and Kelly, and everybody."

"Whose gang am I in?"

"Whitey Miller's."

Whitey Miller, in the opinion of nearly every bridgeman who had ever worked within a mile of him, was the toughest, meanest, pushiest pusher on the Verrazano-Narrows Bridge. Drilling swallowed hard.

CHAPTER EIGHT

THE INDIANS

That night in Johnny's Bar near the bridge, the men spoke about little else.

"Hear about Drilling?"

"Yeah, poor guy."

"They put him in Whitey Miller's gang."

"It's a shame."

"That Whitey Miller's a great ironworker, though," one cut in. "You gotta admit that."

"Yeah, I admit it, but he don't give a crap if you get killed."

"I wouldn't say that."

"Well, I would. I mean, he won't even go to your goddam funeral, that Whitey Miller."

But in another bar in Brooklyn that night, a bar also filled with iron-workers, there was no gloom—no worries about Whitey Miller, no anguish over Drilling. This bar, The Wigwam, at 75 Nevins Street in the North Gowanus section, was a few miles away from Johnny's. The Wigwam was where the Indians always drank. They seemed the most casual, the most detached of ironworkers; they worked as hard as anybody on the bridge, but once the workday was done they left the bridge behind, forgot all about it, lost it in the cloud of smoke, the bubbles of beer, the jukebox jive of The Wigwam.

This was their home away from home. It was a mail drop, a club. On weekends the Indians drove four hundred miles up to Canada to visit their wives and children on the Caughnawaga reservation, eight miles from Montreal on the south shore of the St. Lawrence River; on weeknights they all gathered in The Wigwam drinking Canadian beer (sometimes as many as twenty bottles apiece) and getting drunk together, and lonely.

On the walls of the bar were painted murals of Indian chiefs, and there also was a big photograph of the Indian athlete Jim Thorpe. Above the entrance to the bar hung a sign reading: THE GREATEST IRON WORKERS IN THE WORLD PASS THRU THESE DOORS.

The bar was run by Irene and Manuel Vilis—Irene being a friendly, well-built Indian girl born into an ironworkers' family on the Caughnawaga reservation; Manuel, her husband, was a Spanish card shark with a thin upturned mustache; he resembled Salvadore Dali. He was born in Galicia, and after several years in the merchant marine, he jumped ship and settled in New York, working as a busboy and bottle washer in some highly unrecommended restaurants.

During World War II he joined the United States Army, landed at Normandy, and made a lot of money playing cards. He saved a few thousand dollars this way, and, upon his discharge, and after a few years as a bartender in Brooklyn, he bought his own saloon, married Irene, and called his place The Wigwam. More than seven hundred Indians lived within ten blocks of The Wigwam, nearly every one of them ironworkers. Their fathers and grandfathers also were ironworkers. It all started on the Caughnawaga reservation in 1886, when the Dominion Bridge Company began constructing a cantilever bridge across the St. Lawrence River. The bridge was to be built for the Canadian Pacific Railroad, and part of the construction was to be on Indian property. In order to get permission to trespass, the railroad company made a deal with the chiefs to employ Indian labor wherever possible. Prior to this, the Caughnawagas—a tribe of mixed-blood Mohawks who had always rejected Jesuit efforts to turn them into farmers—earned their living as canoemen for French fur traders, as raft riders for timbermen, as traveling circus performers, anything that would keep them outdoors and on the move, and that would offer a little excitement.

When the bridge company arrived, it employed Indian men to help the ironworkers on the ground, to carry buckets of bolts here and there, but not to risk their lives on the bridge. Yet when the ironworkers were not watching, the Indians would go walking casually across the narrow beams as if they had been doing it all their lives; at high altitudes the Indians seemed, according to one official, to be "as agile as goats." They also were eager to learn the bridge business—it offered good pay, lots of travel—and within a year or two, several of them became riveters and connectors. Within the next twenty years dozens of Caughnawagas were working on bridges all over Canada.

In 1907, on August 29, during the erection of the Quebec Bridge over the St. Lawrence River, the span collapsed. Eighty-six workmen, many Caughnawagas, fell, and seventy-five ironworkers died. (Among the engineers who investigated the collapse, and concluded that the designers were insufficiently informed about the stress capability of such large bridges, was O. H. Ammann, then twenty-seven years old.)

The Quebec Bridge disaster, it was assumed, would certainly keep the Indians out of the business in the future. But it did just the opposite. The disaster gave status to the bridgeman's job—accentuated the derring-do that Indians previously had not thought much about—and consequently more Indians became attracted to bridges than ever before.

During the bridge and skyscraper boom in the New York metropolitan area in the twenties and thirties, Indians came down to New York in great numbers, and they worked, among other places, on the Empire State Building, the R.C.A. Building, the George Washington Bridge, Pulaski Skyway, the Waldorf-Astoria, Triborough Bridge, Bayonne Bridge, and Henry Hudson Bridge. There was so much work in the New York City area that Indians began renting apartments or furnished rooms in the North Gowanus section of Brooklyn, a centralized spot from which to spring in any direction.

And now, in The Wigwam bar on this Friday night, pay night, these grandsons of Indians who died in 1907 on the Quebec Bridge, these sons of Indians who worked on the George Washington Bridge and Empire State, these men who were now working on the biggest bridge of all, were not thinking much about bridges or disasters; they were thinking mostly about home, and were drinking Canadian beer, and listening to the music.

"Oh, these Indians are crazy people," Manuel Vilis was saying, as he sat in a corner and shook his head at the crowded bar. "All they do when they're away from the reservation is build bridges and drink."

"Indians don't drink any more than other ironworkers, Manny," said Irene sharply, defending the Indians, as always, against her husband's criticism.

"The hell they don't," he said. "And in about a half-hour from now, half these guys in here will be loaded, and then they'll get in their cars and drive all the way up to Canada."

They did this every Friday night, he said, and when they arrived on the reservation, at 2 A.M., they all would honk their horns, waking everybody up, and soon the lights would be on in all the houses and everybody would be drinking and celebrating all night—the hunters were home, and they had brought back the meat.

Then on Sunday night, Manuel Vilis said, they all would start back to New York, speeding all the way, and many more Indians would die from automobile accidents along the road than would ever trip off a bridge. As he spoke, the Indians continued to drink, and there were ten-dollar and twenty-dollar bills all over the bar. Then, at 6:30 P.M., one Indian yelled to another, "Com'on, Danny, drink up, let's move." So Danny Montour, who was about to drive himself and two other Indians up to the reservation that night, tossed down his drink, waved goodbye to Irene and Manuel, and prepared for the four-hundred-mile journey.

Montour was a very handsome young man of twenty-six. He had blue eyes, sharp, very un-Indian facial features, almost blond hair. He was married to an extraordinary Indian beauty and had a two-year-old son, and each weekend Danny Montour drove up to the reservation to visit them. He had named his young son after his father, Mark, an ironworker who had crippled himself severely in an automobile accident and had died not long afterward. Danny's paternal grandfather had fallen with the Quebec Bridge in 1907, dying as a result of injuries. His maternal grandfather, also an ironworker, was drunk on the day of the Quebec disaster and, therefore, in no condition to climb the bridge. He later died in an automobile accident.

Despite all this, Danny Montour, as a boy growing up, never doubted that he would become an ironworker. What else would bring such money and position on the reservation? To not become an ironworker was to become a farmer—and to be awakened at 2 A.M. by the automobile horns of returning ironworkers.

So, of the two thousand men on the reservation, few became farmers or clerks or gas pumpers, and fewer became doctors or lawyers, but 1,700 became ironworkers. They could not escape it. It got them when they were babies awakened in their cribs by the horns. The lights would go on, and their mothers would pick them up and bring them downstairs to their fathers, all smiling and full of money and smelling of whiskey or beer, and so happy to be home. They were incapable of enforcing discipline, only capable of handing dollar bills around for the children to play with, and all Indian children grew up with money in their hands. They liked the feel of it, later wanted more of it, fast—for fast cars, fast living, fast trips back and forth between long weekends and endless bridges.

"It's a good life," Danny Montour was trying to explain, driving his car up the Henry Hudson Parkway in New York, past the George Washington Bridge. "You can see the job, can see it shape up from a hole in the ground to a tall building or a big bridge."

He paused for a moment, then, looking through the side window at the New York skyline, he said, "You know, I have a name for this town. I don't know if anybody said it before, but I call this town the City of Man-made Mountains. And we're all part of it, and it gives you a good feeling—you're a kind of mountain builder . . ."

"That's right, Danny-boy, old kid," said Del Stacey, the

Indian ironworker who was a little drunk, and sat in the front seat next to Montour with a half-case of beer and bag of ice under his feet. Stacey was a short, plump, copper-skinned young man wearing a straw hat with a red feather in it; when he wanted to open a bottle of beer, he removed the cap with his teeth.

"Sometimes though," Montour continued, "I'd like to stay home more, and see more of my wife and kid . . ."

"But we can't, Danny-boy," Stacey cut in, cheerfully. "We gotta build them mountains, Danny-boy, and let them women stay home alone, so they'll miss us and won't get a big head, right?" Stacey finished his bottle of beer, then opened a second one with his teeth. The third Indian, in the back seat, was quietly sleeping, having passed out.

Once Montour had gotten the car on the New York Thruway, he began to speed, and occasionally the speedometer would tip between ninety and one hundred miles per hour. He had had three or four drinks at The Wigwam, and now, in his right hand, he was sipping a gin that Stacey had handed him; but he seemed sober and alert, and the expressway was empty, and every few moments his eyes would peer into the rear-view mirror to make certain no police car was following.

Only once during the long trip did Montour stop; in Malden, New York, he stopped at a Hot Shoppe for ten minutes to get a cup of coffee—and there he saw Mike Tarbell and several other Indians also bound for Canada. By 11 P.M. he was speeding past Warrensburg, New York, and an hour or so later he had pulled off the expressway and was on Route 9, a two-lane backroad, and Stacey was yelling, "Only forty miles more to go, Danny-boy."

Now, with no radar and no cars coming or going, Danny Montour's big Buick was blazing along at 120 miles per hour, swishing past the tips of trees, skimming over the black road—and it seemed, at any second, that a big truck would surely appear in the windshield, as trucks always appear, suddenly, in motion-picture films to demolish a few actors near the end of the script.

But, on this particular night, there were no trucks for Danny Montour.

At 1:35 A.M., he took a sharp turn onto a long dirt road, then sped past a large black bridge that was silhouetted in the moonlight over the St. Lawrence Seaway—it was the Canadian railroad bridge that had been built in 1886, the one that got Indians started as ironworkers. With a screech of his brakes, Montour stopped in front of a white house.

"We're home, you lucky Indians," he yelled. The Indian in the back seat who had been sleeping all this time woke up, blinked. Then the lights went on in the white house; it was Montour's house, and everybody went in for a quick drink, and soon Danny's wife, Lorraine, was downstairs, and so was the two-year-old boy, Mark. Outside, other horns were honking, other lights burned; and they remained alive, some of them, until 4 A.M. Then, one by one, they went out, and the last of the Indians fell asleep—not rising again until Saturday afternoon, when they would be awakened, probably, by the almost endless line of bill collectors knocking on doors: milkmen, laundrymen, newsboys, plumbers, venders of vacuum cleaners, encyclopedia salesmen, junk dealers, insurance salesmen. They all waited until Saturday afternoon, when the ironworker was home, relaxed and happy, to separate him from his cash.

The reservation itself is quiet and peaceful. It consists of a two-lane tar road that curves for eight miles near the south shore of the St. Lawrence River. Lined along both sides of the road and behind it are hundreds of small white frame houses, most of them with porches in front—porches often occupied by old Indian men. They slump in rocking chairs, puffing pipes, and quietly watch the cars pass or the big ships float slowly through the St. Lawrence Seaway—ships with sailors on deck who wave at any Indian women they see walking along the road.

Many of the younger Indian women are very pretty. They buy their clothes in Montreal shops, have their hair done on Friday afternoons. There is little about their style of clothing or about their homes that is peculiarly Indian—no papooses, no totems, no Indian gadgets on walls. Some Indian homes do not have running water, and have outhouses in the back, but all seem to have television sets. The only sounds heard on Saturday afternoon are the clanging bells of the Roman Catholic church along the road—most Indians are Catholics—and occasionally the honking of an Indian motorcade celebrating a wedding or christening.

The only road signs bearing Indian symbols are CHARLIE MOHAWK'S SNACK BAR and, on the other side of the road up a bit, the CHIEF POKING FIRE INDIAN MUSEUM. The snack bar's sign is hung principally to amuse the tourists who pass in a yellow bus each day; inside, however, the place looks like any soda fountain, with the booths cluttered with teenaged boys sporting ducktail haircuts and smoking cigarettes, and teenaged girls in skin-tight dungarees and ponytails, all of them twisting or kicking to the rock 'n' roll music blaring from the jukebox.

At Chief Poking Fire's museum, things are different; here it's strictly for the tourists, with the Chief and his family assembling in full regalia a few times each day to dance, whoop, and holler for the tourists and wave tomahawks so that the tourists, clicking their 16-mm. cameras, will have something to show for their visit to an Indian reservation.

The Indian mayor of Caughnawaga is John Lazare, who believes he might be a Jewish Indian. He succeeded his brother, Matthew, as mayor, and Matthew succeeded their father. The Lazares run a gas station on the same side of the road as Chief Poking Fire's museum, and they also sell liquid gas to Indians for home use.

The political viewpoint that has kept the Lazares popular with other Indians all these years is Mayor Lazare's speeches that usually include the sentence, "The Indian should be allowed to do whatever he wants," and also the Lazares' long-time denunciation of the license plate on automobiles. Indians hate to drive with license plates on their cars and would like to remove them, presumably so they'll get fewer speeding tickets (although many Indians ignore all tickets on the grounds that they are not valid documents, having never been agreed to by treaty).

On Saturday afternoons, when the Indian men get out of bed (if they get out of bed), they usually play lacrosse, if it is not too cold. In summer months they might spend their afternoons skimming along the St. Lawrence Seaway in a motorboat they themselves built, or fishing or watching television. On Sunday morning they have their traditional breakfast of steak and cornbread, and usually loll around the house all morning and visit friends in the afternoon.

Then, anywhere from 8 P.M. to 11 P.M., the big cars filled with ironworkers will begin to rumble down the reservation's roads, and then toward the routes to the expressway back to New York. It is a sad time for Indian women, these Sunday evenings, and the ride back to New York seems twice as long to the men as did the Friday-night ride coming up. The alcohol that many of them sip all the way back to New York is the only thing that helps make the trip endurable—and the thing that may help kill them.

And so on this Sunday evening, Danny Montour kissed his

wife goodbye, and hugged his son, and then went to pick up the others for the long ride back.

"Now be careful," Lorraine said from the porch.

"Don't worry," he said.

And all day Monday she, and other Indian women, half-waited for the phone calls, hoping they would never come. And when they did not come on this particular Monday, the women were happy, and by midweek the happiness would grow into a blithe anticipation of what was ahead—the late-Friday sounds of the horns, the croaking call of Cadillacs and Buicks and Oldsmobiles, the sounds that would bring their husbands home . . . and will take their sons away.

CHAPTER NINE

BACK TO
BAY RIDGE

In the spring of 1964, to the astonishment of nobody in this neighborhood that had long suspected it, there was discovered behind
the black curtains and awning and white brick wall at 125 Eighty-
sixth Street, in the plush Colonial Road section of Bay Ridge, a
whorehouse.

Some people, of course, blamed the boomers, recalling the
sight of those slinky blonds who lingered along the shore behind
the bridge. But the *Brooklyn Spectator*, which broke the story on
March 20, after the police finally had sufficient evidence to make
arrests, reported that there were some prominent Bay Ridge citizens among the clientele, although it gave no names. The story

caused a sensation—"the first story of its kind to appear in this paper in its thirty-two-year history," announced the *Spectator*—and
not only was every copy suddenly sold out, but the newspaper office
was left with none for its files, and it hastily had to announce that it
would repurchase, at the regular price of ten cents, any copies of
the March 20 issue in good condition.

After arresting a thirty-six-year-old blond madam who
swore she was a "real estate broker," and two other blond women
who gave their occupations as "baby nurse" and "hostess," the police revealed that even the kitchen of the house had been converted
into a boudoir, that the wallpaper was "vivid," and that there were
mirrors on the ceilings.

Many respectable, old-time Bay Ridge residents were
shocked by the disclosure, and there was the familiar lament for
yesteryear. And a few people, apprehensively gazing up at the al-
most-finished bridge, predicted that soon the bridge might bring
many more changes for the worse—more traffic through residential streets, more and cheaper apartment houses (that might be
crowded with Negroes), and more commercialism in neighborhoods traditionally occupied by two-family houses.

It had been five years since the bridge first invaded Bay
Ridge, and, though the protestors were now quiet and the eight
hundred buildings that stood in the path of the bridge's approachways had now all disappeared, many people had long memories,
and they still hated the bridge.

Monsignor Edward J. Sweeney, whose parish at St. Eph-
rem's had lost two thousand of its twelve thousand parishioners,
thus diminishing the Sunday collection considerably, still became
enraged at the mere mention of the bridge. The dentist Henry

Amen, who had put on forty pounds in the last five years, and was now prosperous in a new office one mile north of his old office, was nevertheless still seething, saying, "I strongly resent the idea of being forced to move."

In some cases the anger in Bay Ridge was as alive in 1964 as it was back in 1959 when "Save Bay Ridge" banners flew; when people screamed "That bridge—who needs it?"; when an undertaker, Joseph V. Sessa, claimed he would lose 2,500 people "from which to draw"; and when the antibridge faction included the disparate likes of housewives, bartenders, a tugboat skipper, doctors, lawyers, a family of seventeen children (two dogs and a cat), a retired prizefighter, a former Ziegfeld Follies girl, two illicit lovers, and hundreds of others who reacted generally as people might react anywhere if, suddenly, the order was delivered: "Abandon your homes—we must build a bridge."

In all, it had taken eighteen months to move out the seven thousand people, and now, in 1964, though a majority of them had been relocated in Bay Ridge, they had lost touch with most of their old neighbors, and had nothing in common now but the memories.

"Oh, those were depressing days," recalled Bessie Gros Dempsey, the former Follies girl who now lives four blocks from the spot where her old home had stood. "When those demolition men moved into the neighborhood, you'd have flower pots full of dust on your windowsills at night, and all day long you'd see them smashing down those lovely homes across the street.

"That crane was like the jaw of a monster, and when it cracked into those buildings, into the roof and ceiling and shingles, everything would turn into powder, and then the dogs would start barking because of all the strange sounds a building makes when it is falling.

"I remember back of where I used to live was this big brownstone—an artist lived there, and the place was built like an Irish castle. When the crane hit into it, it was a horrible sound I'll never forget. And I remember watching them tear down that colonial house that was directly across the street from me. It had columns in front, and a screened-in porch, and it was lived in by a nice elderly couple that had twin daughters, and also an uncle, Jack, a crippled fellow who used to trim those hedges. Such pride was in that home, and what a pity to see that crane smash it all down."

The couple with the twin daughters now lives in upstate New York, Mrs. Dempsey said, adding that she does not know what became of the crippled uncle named Jack. The artist who lived in the brownstone behind her old home is now dead, she said, along with five other people she used to know in her neighborhood in the prebridge days.

Mrs. Dempsey and many others in Bay Ridge in 1964 were citing the bridge as an accomplice in the death of many residents of the old neighborhood; they said that the tension and frustration in losing one's home and the uncertainty of the future had all contributed to the death of many since 1959. One woman pointed out that her husband, never ill before, suddenly had a heart attack and died after a "Save Bay Ridge" rally, and another woman blamed the bridge for her faltering eyesight, saying she never had to wear glasses until the announcement that her home would be destroyed by "that bridge."

Most of the older people who had owned their homes, particularly those on pensions or small fixed incomes, said that the relocation caused them financial hardships because they could not match the price of a new home of comparable size in a comparable neighborhood.

There were, to be sure, a minority who said they were happy that the bridge had forced them to move, or who felt that they had been unjustifiably pessimistic about the changes the Verrazano-Narrows Bridge would bring. Mrs. Carroll L. Christiansen, who had moved from Bay Ridge to Tenafly, New Jersey, into a suburban home with a quarter-acre of land around it, said, "It's a lot better here than in Brooklyn." She added, "In Brooklyn the people didn't mix socially—and never had too much to do with one another. But here it is entirely different. I've learned to play golf since coming here. And my husband and I play cards with other couples in the evenings, and we go to dances at the country club. My daughter, who is seventeen now, felt uprooted for about a year or so, but since then she's also made lots of new friends and the whole life is much easier here."

The undertaker, Joseph Sessa, who had feared he would lose thousands of people, was surviving nicely in Bay Ridge five years later; and the two lovers—the divorced man and the unhappily married woman who used to live across the street from him— have gone their separate ways (she to Long Island, he to Manhattan) and neither blames the bridge for coming between them. "It was just a passing fancy," she says of her old affair, now being moderately contented with her new home, her husband, and children. The lover, a forty-six-year-old insurance company executive, has met a girl at the office, unmarried and in her middle thirties, and each evening they meet in a dimly lit cocktail lounge on Park Avenue South.

Florence Campbell, the divorcee who with her young son had held out in her old apartment until 1960, despite the murder on the floor below, now believes the bridge has changed her life for the better. In her new block, she was introduced by a friend to a merchant mariner, and a year later they were married and now live in a comfortable home on Shore Road.

The old shoemaker who had screamed "sonamabitch" at the bridge authorities for tearing down his little store five years ago, and who returned disillusioned to Cosenza in Southern Italy, has since come back to Brooklyn and is working in another shoe store. He became restless in Italy and found life among his relatives unbearable.

Mr. and Mrs. John G. Herbert, parents of seventeen children, all of whom once lived in a noisy and tattered frame house on the corner of Sixty-seventh and Seventh Avenue in Bay Ridge before the bridge intruded, now live on Fifty-second Street in a three-story, nine-room house that they own, and, in a sense, they are better off than when they were only renting at the old place.

This newer house is two rooms larger than the other one, but it is not any more spacious, and it is also jammed in the middle of a block of teeming row houses. The Herbert children miss the rambling grass yard and trees that used to surround the old property.

Mr. Herbert, a short, muscular Navy Yard worker with blue eyes and a white crew cut, sometimes escapes the clatter and confusion of his home by drinking heavily, and when guests arrive he often greets them by pounding them on the back, pouring them a drink, and shouting, "Com'on, relax—take off your coat, sit down, have a drink, *relax*," and Mrs. Herbert, shaking her head sadly, half

Verrazano-Narrows Bridge Construction
Steel erection completed for upper and lower decks

moans to the guests, "Oh, you're lucky you don't live here," and then Mr. Herbert, downing another drink, pounds the guests again and repeats, "Com'on, relax, have another drink, relax!"

Two of the Herbert boys—Eugene, who is twenty, and Roy, who is nineteen—are very sensitive to such scenes, and both recall how happy, how hopeful they'd been five years before when they had first heard that their old house would be torn down. Finally, they thought, they'd be out of the city altogether, and moving into the country as their father had so often said they would.

When this did not happen, the family being unable to afford any home except the one they now have, the boys felt a bit
cheated; even five years afterward, they missed their old home,
yearned for another like it. One day in the early spring of 1964,
Eugene and Roy took a nostalgic journey back to their old neighborhood, a mile and a half away, and revisited the land upon which
their old home had stood.

Now all was flattened and smoothed by concrete—it was
buried by the highway leading to the bridge, the path toward the
tollgates. The highway was three months away from completion,
and so it was without automobiles. It was quiet and eerie. Eugene
walked around in the middle of the empty highway and then
stopped and said, "It was about here, Roy—this is where the house
was."

"Yeah, I guess you're right," Roy said, "because over there's
the telephone pole we used to climb . . ."

"And over here was where the porch was . . ."

"Yeah, and remember how we used to sit out there at night
in the summertime with the radio plugged in, and remember when
I'd be on that swinging sofa at night with Vera?"

"Boy, I remember that Vera. What a build!"

"And remember when on Friday nights we all used to sit on
the steps waiting for Dad to come home from the Navy Yard with a
half-gallon of ice cream?"

"I remember, and he never failed us, did he?"

"Nope, and I remember what we used to sing, all of us kids,
as we waited for him . . . You remember?"

"Yeah," said Roy. Then both of them, in chorus, repeated
their familiar childhood song:

You scream, I scream,
We all scream
For ice cream.

You scream, I scream,
We all scream
For ice . . .

They looked at one another, a little embarrassed, then re-mained quiet for a moment. Then they walked away from where the house had stood, crossed the empty highway, and, turning around slowly, they rediscovered, one by one, other familiar sights. There was the sidewalk upon which they used to roller-skate, the cement cracks as they had remembered them. There were some of the homes that had not been destroyed by the new highway. There was Leif Ericson Park, where, as boys, they played, and where they once had dug a deep hole in the grass within which to bury things—Scout knives, rings, toys, new baseballs—anything that they had wanted to keep away from their brothers and sisters, because at home nothing was private, nobody respected another's ownership.

They searched along the grass for the hole that they had covered with a metal plate, but could not find it. Then they crossed the street to one of the few houses left on the block, and an elderly woman was shaking a mop outside of a window, and Eugene called up to her, "Hello, Mrs. Johnson, we're the Herbert boys. Remember, we used to live across the street?"

"Why, yes," she said, smiling. "Hardly would have recog-nized you. How are you?"

"Fine. We're over on Fifty-second Street now."

"Oh," she said, softly. "And how's your mother?"

"Fine, Mrs. Johnson."

"Well, give her my regards," the woman said, smiling, then she pulled the mop in and closed the window.

The boys walked on through the vacated neighborhood, past the yellow bulldozers and cement mixers that were quiet on this Saturday afternoon; past the long dirt road that would soon be paved; past the places that had once been alive with part of them.

"Roy, remember that barking dog that used to scare hell out of us?"

"Yeah."

"And remember that candy store that used to be here?"

"Yeah, Harry's. We used to steal him blind."

"And remember . . ."

"Hey," Roy said, "I wonder if Vera is still around?"

"Let's get to a phone booth and look her up."

They walked three blocks to the nearest sidewalk booth, and Roy looked up the name and then called out, "Hey, here it is— SHore 5-8486."

He put in a dime, dialed the number, and waited, thinking how he would begin. But in another second he realized there was no need to think any longer, because there was only a click, and then the coldly proper voice of a telephone operator began, "I am sorry. The number you have reached is not in service at this time. . . . This is a recording."

Roy picked out the dime, put it in his pocket. Then he and his brother walked quietly to the corner, and began to wait for the bus— but it never came. And so, without saying anything more, they began to walk back to their other home, the noisy one, on Fifty-second

Street. It was not a long walk back—just a mile and a half—and yet in 1959, when they were young teenagers, and when it had taken the family sixteen hours to move all the furniture, the trip to the new house in a new neighborhood had seemed such a voyage, such an adventure.

Now they could see, as they walked, that it had been merely a short trip that had changed nothing, for better or for worse—it was as if they had never moved at all.

RAMBLIN' FEVER

A disease common among ironworkers—an itchy sensation called "ramblin' fever"—seemed to vibrate through the long steel cables of the bridge in the spring of 1964, causing a restlessness, an impatience, a tingling tension within the men, and many began to wonder: "Where next?"

Suddenly, the bridge seemed finished. It was not finished, of course—eight months of work remained—but all the heavy steel units were now linked across the sky, the most dangerous part was done, the challenge was dying, the pessimism and cold wind of winter had, with spring, been swept away by a strange sense of surety that nothing could go wrong: a punk named Roberts slipped off the bridge, fell toward the sea—and was caught in a net; a heavy drill was dropped and sailed down directly toward the scalp of an Indian named Joe Tworivers—but it nipped only his toes, and he grunted and kept walking.

The sight of the sixty-four-hundred-ton units all hanging horizontally from the cables, forming a lovely rainbow of red steel across the sea from Staten Island to Brooklyn, was inspiring to spectators along the shore, but to the ironworkers on the bridge it was a sign that boredom was ahead. For the next phase of construction, referred to in the trade as "second-pass steel," would consist primarily of recrossing the entire span while lifting and inserting small pieces of steel into the structure—struts, grills, frills—and then tightening and retightening the bolts. When the whole span had been filled in with the finishing steel, and when all the bolts had been retightened, the concrete mixers would move in to pave the roadway, and next would come the electricians to string up the lights, and next the painters to cover the red steel with coats of silver.

And finally, when all was done, and months after the last ironworker had left the scene for a challenge elsewhere, the bridge would be opened, bands would march across, ribbons would be cut, pretty girls would smile from floats, politicians would make speeches, everyone would applaud—and the engineers would take all the bows.

And the ironworker would not give a damn. He will do his boasting in bars. And anyway, he will know what he has done, and he would somehow not feel comfortable standing still on a bridge, wearing a coat and tie, showing sentimentality.

In fact, for a long time afterward he will probably not even think much about the bridge. But then, maybe four or five years later, a sort of ramblin' fever in reverse will grip him. It might occur while he is driving to another job or driving off to a vacation; but suddenly it will dawn on him that a hundred or so miles away stands one of his old boom towns and bridges. He will stray from his course, and soon he will be back for a brief visit: Maybe it is St. Ignace, and he is gazing up at the Mackinac Bridge; or perhaps it's San Francisco, and he is admiring the Golden Gate; or perhaps (some years from now) he will be back in Brooklyn staring across the sea at the Verrazano-Narrows Bridge.

Today he will doubt the possibility, most of the boomers will, but by 1968 or '69 he probably will have done it: He may be in his big car coming down from Long Island or up from Manhattan, and he will be moving swiftly with all the other cars on the Belt Parkway, but then, as they approach Bay Ridge, he will slow up a bit and hold his breath as he sees, stretching across his windshield, the Verrazano—its span now busy and alive with auto traffic, bumper-to-bumper, and nobody standing on the cables now but a few birds.

Then he will cut his car toward the right lane, pulling slowly off the Belt Parkway into the shoulder, kicking up dust, and motorists in the cars behind will yell out the window, "Hey, you idiot, watch where you're driving," and a woman may nudge her husband and say, "Look out, dear, the man in that car looks drunk."

In a way, she will be right. The boomer, for a few moments, will be under a hazy, heady influence as he takes it all in—the sights and sounds of the bridge he remembers—hearing again the rattling and clanging and Hard Nose Murphy's angry voice; and remembering, too, the cable-spinning and the lifting and Kelly saying, "Up on seven, easy now, easy"; and seeing again the spot where Gerard McKee fell, and where the clamps slipped, and where the one-thousand-pound casting was dropped; and he will know that on the bottom of the sea lies a treasury of rusty rivets and tools.

The boomer will watch silently for a few moments, sitting in his car, and then he will press the gas pedal and get back on the road, joining the other cars, soon getting lost in the line, and nobody will ever know that the man in this big car one day had knocked one thousand rivets into that bridge, or had helped lift four hundred tons of steel, or that his name is Tatum, or Olson, or Iannielli, or Jacklets, or maybe Hard Nose Murphy himself.

Anyway, this is how, in the spring of 1964, Bob Anderson felt—he was a victim of ramblin' fever. He was itchy to leave the Verrazano job in Brooklyn and get to Portugal, where he was going to work on a big new suspension span across the Tagus River.

"Oh, we're gonna have a ball in Portugal," Anderson was telling the other ironworkers on his last working day on the Verrazano-Narrows Bridge. "The country is absolutely beautiful, we'll have weekends in Paris . . . you guys gotta come over and join me."

"We will, Bob, in about a month," one said.

"Yeah, Bob, I'll sure be there," said another. "This job is finished as far as I'm concerned, and I got to get the hell out. . . ."

On Friday, June 19, Bob Anderson shook hands with dozens of men on the Verrazano and gave them his address in Portugal, and that night many of them joined him for a farewell drink at the Tamaqua Bar in Brooklyn. There were about fifty ironworkers there by 10 P.M. They gathered around four big tables in the back of the room, drank whiskey with beer chasers, and wished Anderson well. Ace Cowan was there, and so was John Drilling (he had just been promoted back to pusher after three hard months in Whitey Miller's gang), and so were several other boomers who had worked with Anderson on the Mackinac Bridge in Michigan between 1955 and 1957.

Everyone was very cheerful that night. They toasted Anderson, slapping him on the back endlessly, and they cheered when he promised them a big welcoming binge in Portugal. There were reminiscing and joke-telling, and they all remembered with joy the incidents that had most infuriated Hard Nose Murphy, and they recalled, too, some of the merry moments they had shared nearly a decade ago while working on the Mackinac. The party went on beyond midnight, and then, after a final farewell to Anderson, one by one they staggered out.

Prior to leaving for Lisbon, Bob Anderson, with his wife, Rita, and their two children, packed the car and embarked on a brief trip up to St. Ignace, Michigan. It was in St. Ignace, during the Mackinac Bridge job, that Bob Anderson had met Rita. She still had parents and many friends there, and that was the reason for the trip—that and the fact that Bob Anderson wanted to see again the big bridge upon which he had worked between 1955 and 1957.

A few days later he was standing alone on the shore of St. Ignace, gazing up at the Mackinac Bridge from which he had once come bumping down along a cable, clinging to a disconnected piece of catwalk for 1,800 feet, and he remembered how he'd gotten up after, and how everybody then had said he was the luckiest boomer on the bridge.

He remembered a great many other things, too, as he walked quietly at the river's edge. Then, ten minutes later, he slowly walked back toward his car and drove to his mother-in-law's to join his wife, and later they went for a drink at the Nicolet Hotel bar, which had been boomers' headquarters nine years ago and where he had won that thousand dollars shooting craps in the men's room.

But now, at forty-two years of age, all this was behind him. He was very much in love with Rita, his third wife, and he had finally settled down with her and their two young children. They both looked forward to the job in Portugal—and the possibility of tragedy there never could have occurred to them.

In Portugal while looking for a house, the Andersons stayed at the Tivoli Hotel in Lisbon. Bob Anderson's first visit to the Tagus River Bridge was on June 17. At that time the men were working on the towers, and the big derricks were hoisting up fifty-ton steel sections that would fit into the towers. Anderson apparently was standing on the pier when, as one fifty-ton unit was four feet off the ground, the boom buckled and suddenly the snapped hoisting cable whipped against him with such force that it sent him crashing against the pier, breaking his left shoulder and cracking open his skull, damaging his brain. Nobody saw the accident, and the bridge company could only guess what had happened. Bob Anderson remained unconscious all day and night and two months later he was still in a coma, unable to recognize Rita or to speak. His booming days were over, the doctors said.

When word got back to the Verrazano in Brooklyn, it affected every man on the bridge. Some were too shocked to speak, others swore angrily and bitterly. John Drilling and other boomers rushed off the bridge and called Rita in Portugal, volunteering to fly over. But she assured them there was nothing they could do. Her mother had arrived from St. Ignace and was helping care for the children.

For the boomers, it was a tragic ending to all the exciting time in New York on the world's longest suspension span. They were proud of Bob Anderson. He had been a daring man on the bridge, and a charming man off it. His name would not be mentioned at the Verrazano-Narrows dedication, they knew, because Anderson—and others like him—were known only within the small world of the boomers. But in that world they were giants, they were heroes never lacking in courage or pride—men who remained always true to the boomer's code: going wherever the big job was

> *. . . and lingering only a little while . . .*
> *then off again to another town, another bridge . . .*
> *linking everything but their lives.*

AFTERWORD

It is a sunny New York afternoon in the late summer, and I am standing near the western anchorage of the Verrazano-Narrows Bridge on the water's edge of Staten Island. I have come here to pose for the author's photograph of this book that is my homage to the bridge that I first climbed forty years ago, when it spanned the harbor in a bridled state of juvenility without purpose or responsibility, lacking at the time the concrete roadway, the directional signs, and the toll plaza that would make it functional and self-supporting.

Now it has grown into a commercial colossus that is crossed every day by 250,000 vehicles that deposit a daily sum of one million dollars, and it looms as the gateway to the city's harbor and to the high hopes that guide the entrepreneurial and persevering intrinsicality of New Yorkers. This day the bridge itself was receiving a facelift from crews of maintenance workers who, moving up and down the towers and across the span in cable-suspended metal carts, were scraping rust spots off the surface with wire brushes, and smearing reddish lead waterproofing paste over the steel before applying coats of light gray paint with lamb's-wool rollers attached to long poles. The bridge is scraped and repainted from top to bottom every ten years. It takes five years to complete the job, costing management about seventy-five million dollars.

The general manager of the bridge oversees its day-to-day operation from a brick office off the ramp near the Staten Island anchorage. He is a mild-mannered man of fifty-two who stands six-feet four, weighs 250 pounds, and is named Robert Tozzi. As a college student more than thirty years ago at St. John's University in Brooklyn, he worked for three summers as a toll collector on the Verrazano, beginning in 1969, five years after it opened, when the fare for each passenger vehicle was $.50 (it is now a round-trip fare of $7.00 per passenger vehicle) and amounted to almost one thousand dollars in coins and currency during each eight-hour work shift. As a toll collector Robert Tozzi was paid twenty dollars a day.

The booth he used to occupy is a short walk away from his present-day office. Hanging from his office walls are monitors showing the flow of traffic across the bridge from thirty-two different camera angles—offering views from the two towers, the upper and lower roadways, the anchorages, and other points—but nothing he sees varies much from what he saw the day before, or the day before that. What happens on the bridge is quite predictable to him. Three vehicles will break down on the bridge almost every day, briefly causing traffic jams. He expects, and usually gets, two auto crashes a day—fender benders or other collisions not resulting in serious injuries. Traffic fatalities are rare, perhaps one every two years. The last one occurred in 1999. Each year, on average, two individuals with suicidal tendencies select the Verrazano-Narrows Bridge as the jumping off point to end their lives.

Robert Tozzi's father was a maintenance manager with the Triborough Bridge and Tunnel Authority, which built many of the city's toll-collecting facilities, including the Verrazano, and it was through his father's introduction that the younger Tozzi came to the agency. After three summers in the tollbooth, he was hired full time in 1973 to drive one of the Verrazano's tow trucks, helping to remove disabled cars from the roadway.

In 1974, when he was twenty-four, Tozzi was offered the opportunity to drive one of the agency's sedans in order to chauffeur the Triborough's chairman, Robert Moses. For the next four-and-a-half years, dutifully and diplomatically, Tozzi drove Moses to countless locations. He knew that Moses, deficient in grace even in the best of times, was embittered over the bad press and negative commentary he had been receiving since the publication in 1974 of Robert Caro's highly critical book about him, entitled *The Power Broker* and subtitled: *Robert Moses and the Fall of New York*. Approaching the end of his long career as an urban planner—he had paved the way for the creation of the Verrazano-Narrows Bridge and other great undertakings that he believed the public needed and should appreciate—Moses as an octogenarian found himself being second-guessed and excoriated on talk shows and in the media for his past decisions to demolish vast areas of living space and dispossess multitudes of people in order to build new highways, bridges, tunnels, and tollgates. In 1975 the Caro book won the Pulitzer Prize, adding insult to injury as far as Moses was concerned.

On the day I met Robert Tozzi—he escorted me to the anchorage himself, a necessary courtesy attributable to security policies instituted since the World Trade Center attacks—he said that he had never gotten around to buying *The Power Broker*, although the book is popular with many of his fellow employees.

Robert Moses is now long deceased, as is the Verrazano's chief designer, O. H. Ammann, and also many of the residents and businesspeople of Brooklyn and Staten Island who in the late 1950s tried without success to prevent Moses from destroying their neighborhoods that stood in the proposed pathways to the bridge. Some of them, however, recently told me that their worst fears had failed to transpire, that the new location into which they had been forced to move their businesses or domiciles turned out not to be as disruptive or depressing as they had imagined. The structures they had moved to were generally in better condition than the ones they had surrendered to the wrecking crews; and while what they lost in sentimental value was irredeemable, and while they continued to resent the arbitrary power that Moses had wielded over their lives, they gradually became resigned to the inevitability of change and their inability to resist it.

Among those who spoke out in 1959 against Moses's plan for the bridge was a dentist, Henry Amen, whose office in the Bay Ridge section of Brooklyn was designated to be leveled and paved over by the road builders. Dr. Amen was fortunate in finding a new office convenient enough for his Bay Ridge–based practice; he is still there today, working at the age of seventy-six, having as a partner a dentist of thirty-four who is his daughter.

Another anti-Moses spokesman who I interviewed in the late 1950s was Joseph V. Sessa, who had somberly predicted his firm's demise after learning that the condemned properties in his neighborhood would result in the dispersal of 2,500 families "from which to draw." This figure was more than a third of the estimated

7,000 people who were being dislodged; but the mortician, unlike the dentist, would not see his own workplace destroyed. It was located along the fringe of the targeted area, on Fort Hamilton Parkway, and it remains there as it was, still dedicated to burying people—including Mr. Sessa, who passed away in 1977 at the age of seventy-nine. His resourceful grandson, Joseph J. Sessa, now operates the family business and has expanded it to include two other mortuaries located elsewhere in Brooklyn.

The family that was perhaps the most inconvenienced by the construction of the bridge's approachways was that of a Brooklyn Navy Yard worker, John G. Herbert, who, with his wife, Margaret, and fifteen of their seventeen children, resided until 1959 in a comfortable old ramshackle corner house on a hill adjacent to a park in the Bay Ridge area. Since their dwelling was soon to be pulverized, Mr. and Mrs. Herbert were persistently encouraged by the Triborough's resettlement office to select one of the three spacious houses that was available to them less than two miles away. Although they were unenthusiastic about the houses they were shown—none was near a park, all were located on densely populated blocks—they settled for the one closest to that which they were leaving. It took the couple twelve trips and eighteen hours to convey their children, their furniture, and the rest of their possessions; and, other than the resentment expressed along the way by their fifteen-year-old boy, Eugene, and his fourteen-year-old brother, Roy, the transfer was completed with efficiency.

After Eugene and Roy had been told to walk back to the old homestead to retrieve a pet cat that had been forgotten, they expressed their feelings about the displacement by throwing bricks through what had been their bedroom windows, and with a rusty axe they chopped down the wooden railings of the front porch and the bannisters of the interior staircase, and knocked holes into the plaster walls—a cathartic experience.

More than forty years later, Eugene is fifty-eight and recently retired from his job installing Otis elevators. (His thirteen surviving siblings now reside in various parts of New York, New Jersey, Delaware, Ohio, Minnesota, and Florida.) He lives in a rented second-story Brooklyn apartment with his wife and their twenty-year-old daughter who works in a doctor's office. The apartment is located on a busy residential block lined with row houses and trees and hard-to-find parking spaces. It is in the Bay Ridge section, not far from where Eugene Herbert's parents last lived. Although he pays it little notice, the Brooklyn tower of the Verrazano-Narrows Bridge rises above the rooftops of his neighborhood.

In Staten Island the bridge is also no longer an object of great interest or angst. Here a new generation of settlers, having grown up with it, accept it as essential to their expanding development, the highlight of their horizon, their link to the mammoth mosaic that is Brooklyn and to the seafront towers of Lower Manhattan that the writer Truman Capote once described as a "diamond iceberg."

Although the Staten Island population of 443,000 (in a city of over 8 million) is now nearly double that of the prebridge figure, and while the island's once sprawling farmlands have disappeared along with the country roads through which motorists once moved bumpily past herds of grazing cattle, a provincial atmosphere still prevails here. Along tree-shaded residential streets one sees row upon row of single-family white frame houses with shuttered windows and roadside mailboxes and tidy lawns with poles bearing American flags that were on display long before the recent nationwide proliferation of patriotism prompted by the attacks of terrorists. These flags flew steadfastly in Staten Island as other flags were being burned in Manhattan and elsewhere by college students and others among the anti–Vietnam War activists of the 1960s and early '70s.

Staten Islanders are traditionally conservative, standard-bearing advocates of law and order. A disproportionate number of them serve the city as police officers and firefighters, and among the 343 New York City firefighters who lost their lives in the World Trade Center disaster, 78 were from Staten Island. Residents of Irish ancestry have historically influenced the social and political land-scape here, but now their coreligionists, the Italian-Americans, are the dominant group. A majority of them are related ancestrally to agrarian villages in Southern Italy, and they have reinforced a "vil-lage mentality" on the island, a sense of insularity and regularity, an affinity for familiarity.

This is not to say that Staten Island is without its diversity of newcomers. Asians and Jews from Russia have joined such tenured minorities as Hispanics and African-Americans. But the island re-mains a bastion of blue-collar white families and cadres of middle-management commuters and young women who are employed in Manhattan as secretaries, bank tellers, and sales reps and who, in many cases, will quit these jobs once they marry and have children.

Among the island's commuting population is a strapping, ruddy-complected man of six-feet-two named John McKee, who, after placing his boots, his tool belt, and his hard hat in the trunk of his car, drives to work across the bridge that forty years before he had helped to erect. His two bridge-building brothers were then with him on the job; and on a cloudy, windy Wednesday morning in early October of 1963, eight months before the skeletal steel struc-ture of the Verrazano was fully bracketed and bolted, John McKee's younger brother, Gerard, slipped off one of the cables and fell more than 350 feet into the water, hitting with such force that he immediately lost his life.

There was a work stoppage that day, and subsequently a five-day strike was initiated by the union boss of Local 40, Ray Corbett, who, though fearless when he himself was a young man wearing a hard hat—in 1949 he stood casually atop the Empire State Building overseeing the installation of its 224-foot television tower—believed in 1963, in the wake of Gerard McKee's death and two earlier fatalities, that the Verrazano project was becoming unnecessarily perilous and that safety nets should immediately be strung up under the bridge's framework. This was finally agreed to by the management, and Ray Corbett's concern was justified in the months ahead as three other bridge workers lost their balance but not their lives thanks to the nets.

Among the lucky trio was Robert Walsh, a friend of the McKee family and currently Local 40's business manager. The workers' headquarters building in Manhattan, on Park Avenue South between thirtieth and thirty-first Streets, is named in honor of Ray Corbett, who died in 1992 at the age of seventy-seven.

The two surviving McKee brothers have continued to practice their hazardous craft without debilitating injuries since Gerard's death, but both men are contemplating their retirement. John McKee is fifty-nine. His brother, James, living in Brooklyn, is sixty. James McKee's wife, Nettie, who is an administrative secretary with the borough's school system, has remained in social contact with the late Gerard McKee's onetime fiancée, Margaret Nucito, who now resides in Florida, but in the early 1960s lived across the street from the McKees in Brooklyn, in the working class Catholic neighborhood of Red Hook.

Margaret was then acknowledged to be the prettiest girl on the block, a petite Italian-American redhead who spent her days as a file clerk with the phone company and came directly home after work. She became engaged to Gerard McKee when they were nineteen, after he had dropped out of high school in his final year to enroll in an apprentice program that would in time qualify him to join his older brothers in Local 40.

Margaret and Gerard had been classmates in parochial school and had dated during and after their years in high school, although "dating" in those days in that neighborhood hardly connoted sexual permissiveness. Had Gerard not fallen off the bridge, Margaret would have been his virgin bride, perhaps among the last women of her generation in Red Hook to be so determinedly inclined regarding premarital chastity; and yet along with her firmly held opinions on delayed gratification and the sanctity of marriage, and her feelings of appreciation and tenderness toward Gerard for supporting and respecting her views, she doubted that she would have been very happy as Gerard's wife.

She told me this during an interview in her home seven weeks after Gerard's funeral, which had been attended by hundreds of fellow bridge workers. These men were the *only* individuals that Gerard looked up to, she said; they mattered more to him than any girlfriend or wife ever would. He sought acceptance not only in an all-male world, she suggested, but in a brotherhood of chance takers who bonded together like the steel they connected; and even after their day's work was done, they continued their camaraderie in bars as they talked about the job and exchanged jokes and bragged about themselves and one another, all the while drinking beer and taking their own sweet time about returning to their homes.

Still, had Gerard lived, and despite her reservations about

their compatibility, she said she would have proceeded with their marital plans. It would have been scandalous not to do so. The wedding date had been announced, their union was a fait accompli as far as their kinfolk and friends were concerned. She was a product of a traditional Italian-American family, obedient if at times uncertain. She did not know exactly what she wanted, but she did not want exactly what she had. In 1967, four years after her fiancé's death, Margaret Nucito married a man from outside her neighborhood who in no way reminded her of the late Gerard McKee. He ran a home appliance shop that specialized in selling refrigerators and washing machines, he catered mostly to women, and he was comfortable in their presence.

Another person associated with Gerard McKee that I interviewed during the summer of 2002 was Edward Iannielli, now sixty-seven, but a grieving man of twenty-seven when I first met him in 1963, weeks after he had briefly clung to, and then lost hold of, the two-hundred-pound body of his doomed coworker. Following the funeral, and haunted by the experience of Gerard's death, Edward Iannielli resumed working on the Verrazano-Narrows Bridge until it was completed in 1964. In the years that followed, he was engaged in the construction of about fifty high-rise office buildings in the metropolitan area, and other projects as well, until he decided to retire in 1991. A religious man who regularly attends Mass and believes that everyday events are instilled with a special meaning, he interpreted it as inevitable that he would cap his thirty-six-year career by ending up on the Verrazano, joining forty fellow workers for three months, from the middle of the summer through the fall of 1991 in the task of removing rust from the towers and cables that, along with old memories of sadness, brought him renewed feelings of personal pride and professional achievement.

He was fifty-five years old when he returned to work on the bridge. His back was in pain, he suffered from sciatica, and his left hand had a missing index finger, a permanently bent middle finger, and a fourth finger cut off at the knuckle, all the result of work-related accidents. He wore his faded brown hard hat, his much-traveled tool belt, his blue jeans, and one of the rather gaudy tropical shirts he liked to wear even in cooler weather. He also wore a new, expensive pair of calf-high soft leather boots with rubber soles and no heels, boots he fondly regarded as stumble-proof good-luck boots, the last pair he would ever wear while traversing the beam of a bridge.

Despite his advanced age and his occupational ailments, Edward Iannielli had drawn one of the more difficult assignments on the bridge, that of removing rust from the highest points of the towers. His height (five feet seven inches) and his weight (one hundred and forty pounds) meant that he would not put undue stress on the quarter-inch galvanized cable wire that would carry him more than six hundred feet to the tower tops.

On a particular morning, Iannielli stepped out onto the lower ledge of the tower on the Brooklyn side, overlooking the upper deck of the roadway, two hundred feet above the water, and he eased himself down into a squarish silver metal container that was attached to a cable; with his right hand holding onto the rail of the container, he used his left hand to press the "up" lever that activated the electrical motor lodged in the base of the container, directly beneath his floor space. When he arrived at the top, which took him twenty minutes, his job was to lean out and use wire brushes and scrapers to remove the rust, and then, wearing rubber gloves, to smear a rust-resistant paste onto whatever corrosion existed along the flat surface and bolts of the tower. As he did this he envisioned himself thirty years before inserting those same bolts into the same steel and once more he felt an identity with the great structure. Tears came to his eyes as he continued to work, and, dipping his gloved left hand into a bucket of reddish paste, he reached out to touch an untarnished plate of steel that was secured by a row of bolts and with his bent middle finger he wrote as clearly as he could, in block letters, "Catherine"—the name of his wife of thirty years, who had recently died of cancer.

From his vantage point he could see, extending for miles, the variegated shapes and shades of the city: the verdant parks and tree-lined highways, the smokestacks, church steeples, row houses, apartment buildings, and skyscrapers—the taller they were, the more familiar he was with them.

If helping to build the Verrazano-Narrows Bridge had been the most gratifying job of his life, and it was, then the low point came during the years he was employed as a worker at the World Trade Center. It was never in his nature to be critical of designers and engineers, and particularly so now that the World Trade Center building site has become a memorial shrine. However, during the years he participated in its construction, beginning in 1968 and continuing through 1971, Iannielli and most of the workers he was associated with were appalled by the lightness of the floor beams they were directed to connect, the lack of interior support columns, the seeming fragility of the entire construction, and the hasty pace they were instructed to follow in adding to the skyline of New York two tubular towers that suggested from afar a pair of elongated birdcages.

"Flimsy" was how Iannielli had characterized the formation of the World Trade Center in one of our interviews during the summer of 2002. A day later he telephoned to say he regretted using the word, fearing it made him sound insensitive to the event of September 11. But I reminded him that his feelings about the project's design and stratification had already been expressed by several other workers I had spoken to, many of them Verrazano veterans. I also told him that "flimsy" had been used in a speech delivered months before at Stanford University by Ronald O. Hamburger, a member of a team of structural engineers assessing the performance of the World Trade Center during the terrorist attacks, the ensuing fires, and ultimately the demolition. "The floor trusses were relatively flimsy . . . the trusses just fell apart," Mr. Hamburger said. It was pointed out by other engineers that the World Trade Center buildings were about ninety percent "air," designed to achieve the utmost in rentable floor space and flexibility, unencumbered by columns, explaining why the rubble in the wake of the collapse was only a few stories high. "We didn't find much concrete in there," I was told by one hard-hatted worker who was among the unionists' volunteers who removed the debris. "It was mostly powder, dust, mounds of dust."

In addition to the World Trade Center's design, its building standards, and possible flaws in its methods of fireproofing, Edward Iannielli remembers his working days there as a dispiriting time, an era of conflict in which student antiwar demonstrators and various counterculturists presumed to occupy the moral high ground in New York and elsewhere, while such hard-hatted unionists as himself, patriotic traditionalists opposed to the desecration of the flag, were frequently depicted in the media as reactionary goons and worse.

One day in early May of 1970, Iannielli recalled, a melee broke out near Wall Street between crowds of antiwar activists and dozens of workers who had followed them there. Iannielli had not accompanied his angry coworkers, disinclined to inflict added punishment on his body, but when they returned they told him that they had beaten up many demonstrators and had destroyed countless antiwar banners. They had also stormed City Hall and forced Mayor John Lindsay to raise the flag on the roof to full staff, displeasing the antiwar faction that had earlier convinced him to lower it in memory of the Kent State marchers who had been killed by law enforcement authorities in Ohio earlier in the week.

"When I think of the World Trade Center I think of all the hostility, all the bad feelings that seemed to be built into it from the very beginning," one of Iannielli's coworkers told me. "We who had been on the job were as shocked and depressed as everyone else by what happened to all those innocent people. But as for the buildings, well, we weren't so surprised they went down the way they did."

Some among the five hundred steelworkers who had worked on the World Trade Center had come down to New York from the Indian reservation of mixed-blood Mohawks located along the St. Lawrence River near Montreal; and the worker I most wanted to see again was Danny Montour, an amusing and amiable individual who had befriended me in 1963 after I had met him working on the Verrazano. During this time he had taken me up to the reservation to spend a weekend with his family, introducing me to his wife, Lorraine, and others in his immediate and extended family, all the male members being bridge workers. His father had died on the job in 1956, and his grandfather in 1907. When Danny Montour introduced me to his two-year-old son, Mark, Lorraine, the boy's mother, expressed the hope that he would seek a different means of livelihood than his male kinfolk.

In the summer of 2002, I telephoned the Montour home

and learned from Lorraine that Danny was dead. He had died in 1972 at the age of thirty-four, she said, having fallen ten stories after a concrete ledge had crumbled under him when he was constructing a hospital near JFK Airport in Queens. Their son, Mark, had attended Cornell University for a while, she said, but—"It's in his blood"—he is currently employed with a crew that is erecting a skyscraper in Jersey City that will be completed in 2003.

When I contacted him on his cell phone, Mark Montour explained that he was talking to me while standing on a steel beam seven hundred feet in the air, overlooking Ground Zero on the opposite side of the harbor and with a view of a dozen buildings that his late father helped to build, including the Met Life Building, when it was called the Pan Am, and, of course, the Verrazano-Narrows Bridge, which, even on cloudy days, he can clearly see.

APPENDIX

FACTS ABOUT THE VERRAZANO-NARROWS BRIDGE

TOTAL LENGTH OF BRIDGE (INCLUDING APPROACHES)	13,700 FEET
LENGTH OF SUSPENDED STRUCTURE	6,690 FEET
LENGTH OF MAIN SPAN (LONGEST IN THE WORLD)	4,260 FEET
LENGTH OF EACH SIDE SPAN	1,215 FEET
WIDTH OF BRIDGE	103 FEET
HEIGHT OF TOWERS ABOVE MEAN HIGH WATER	690 FEET
DEEPEST FOUNDATION BELOW MEAN HIGH WATER	170 FEET
CLEARANCE AT CENTER ABOVE MEAN HIGH WATER	228 FEET
NUMBER OF MAIN CABLES	4
LENGTH OF ONE CABLE	7,205 FEET
DIAMETER OF ONE CABLE (COMPACTED)	$35\frac{7}{8}$ INCHES
NUMBER OF STRANDS PER CABLE	61
NUMBER OF WIRES PER STRAND	428
NUMBER OF WIRES PER CABLE	26,108
DIAMETER OF EACH WIRE	0.196 INCH
TOTAL LENGTH OF WIRE IN FOUR CABLES	145,000 MILES
NUMBER OF SUSPENDER ROPES PER CABLE	262
TOTAL WEIGHT OF SUSPENDED STRUCTURE OF BRIDGE	51,000 TONS
STRUCTURAL STEEL IN MAIN BRIDGE	160,000 TONS
CONCRETE IN MAIN BRIDGE	570,000 CUBIC YARDS
NUMBER OF DECKS	2
NUMBER OF TRAFFIC LANES	12

INDEX

INDEX

Waldorf–Astoria, 106

Walkin' boss, 47–48

Wall Street, x, 145

Walsh, Robert, 140

Walt Whitman Bridge, 57, 58

Wasp (carrier), 66

Wigwam, The, 104–5, 106–7, 109

Williamsburg Bridge, 3, 38

Wind, 79–80, 89, 99
  bridge sway in, 41–42

Wire rope and cable, 34, 36, 60, 61

Wives, 2

Women, 2, 6, 8, 109
  Anderson and, 8–9, 10, 11, 62
  in Bay Ridge whorehouse, 115, 116
  Indian, 111, 112, 113

World Trade Center, x, 135, 139, 144–46
  flimsy construction of, 144–45

Wrought iron, 31–32